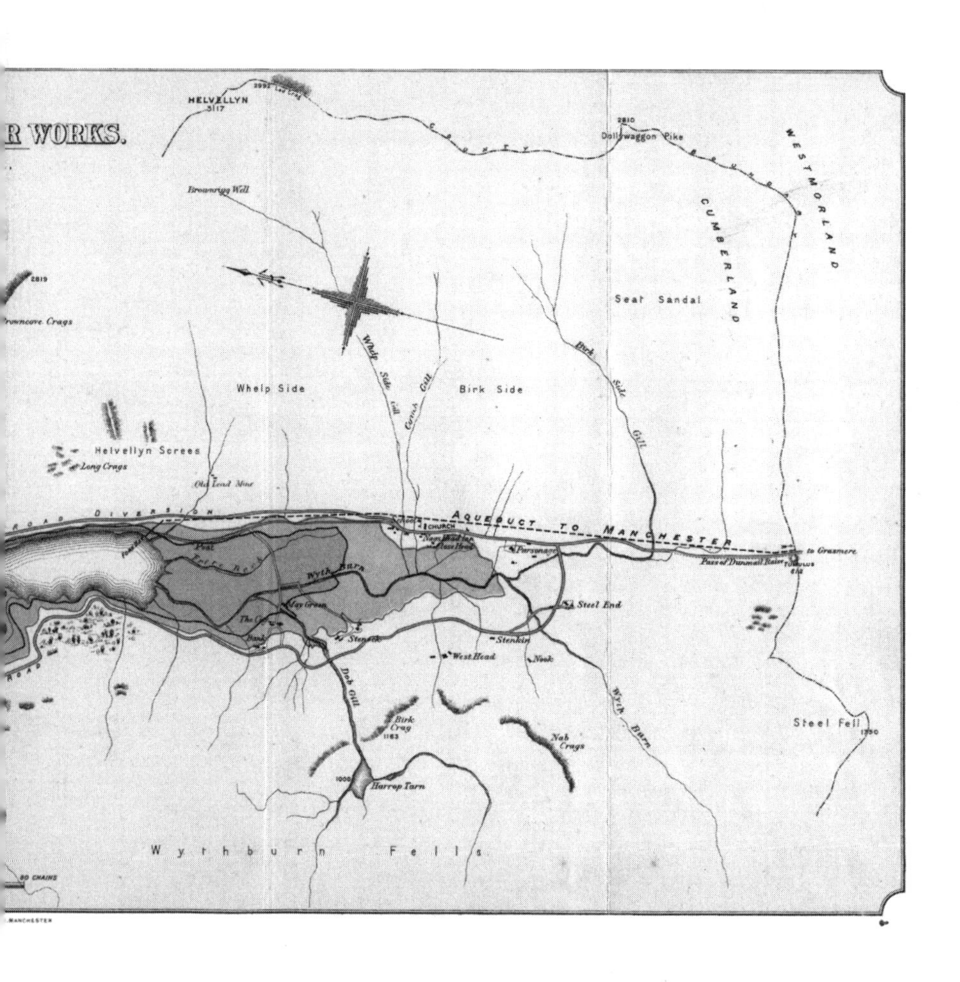
R WORKS.
HELVELLYN
3117
2810
Dollywaggon Pike
WESTMORLAND
CUMBERLAND
Brownrigg Well
Seat Sandal
Whelp Side
Birk Side
Helvellyn Screes
Long Crags
AQUEDUCT TO MANCHESTER
ROAD DIVERSION
CHURCH
Parsonage
Steel End
Pass of Dunmail Raise
to Grasmere
Wyth Burn
Stenkin
West Head
Nook
Dob Gill
Birk Crag
1163
Nab Crags
Harrop Tarn
Steel Fell
Wythburn Fells
80 CHAINS
MANCHESTER

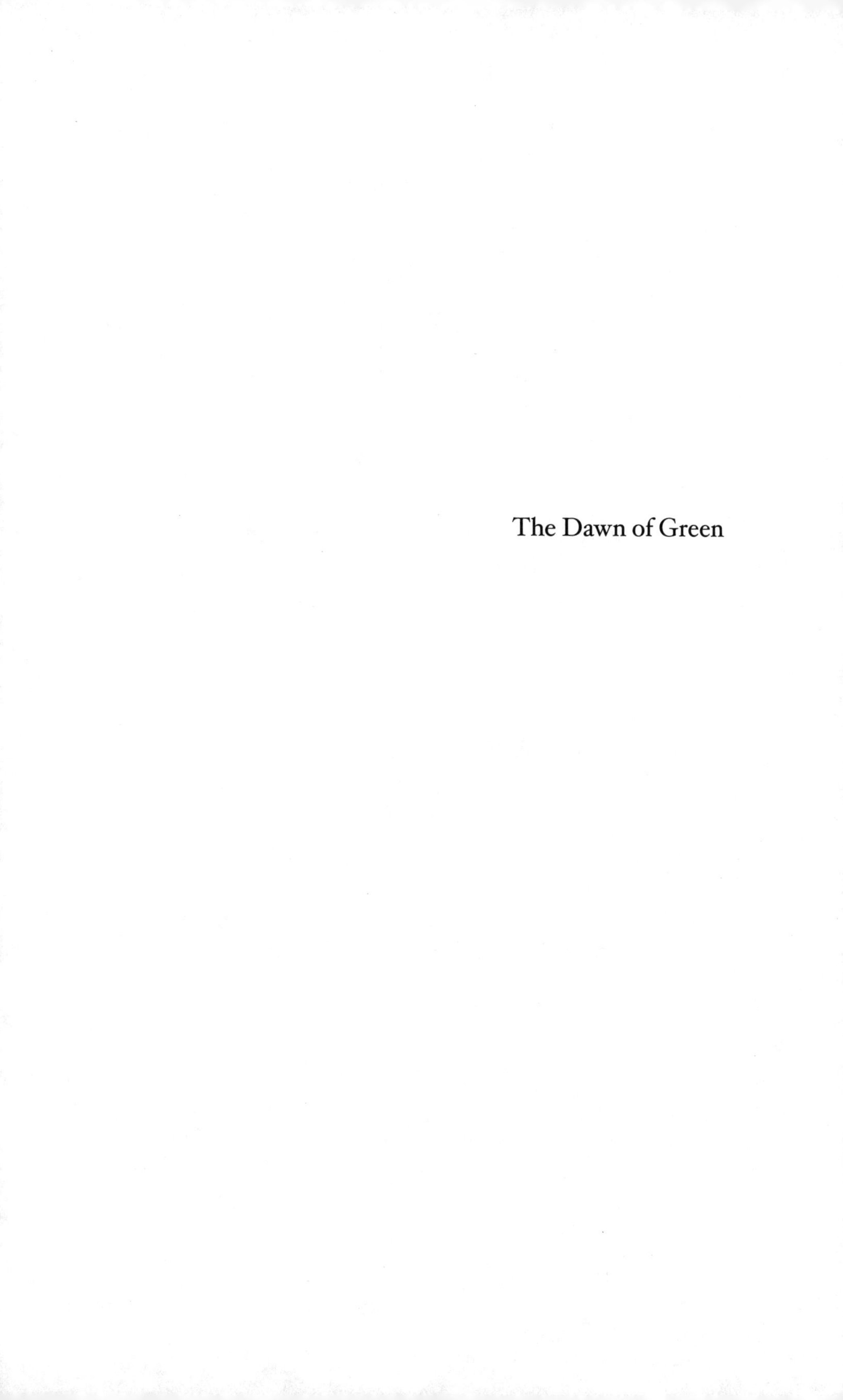

The Dawn of Green

The Dawn of Green

Manchester, Thirlmere, and Modern Environmentalism

HARRIET RITVO

The University of Chicago Press | *Chicago and London*

Harriet Ritvo is the Arthur J. Conner Professor of History at the Massachusetts Institute of Technology and author of *The Platypus and the Mermaid, and Other Figments of Classifying Imagination* and *The Animal Estate: The English and Other Creatures in the Victorian Age.*

The University of Chicago Press, Chicago 60637
The University of Chicago Press, Ltd., London

Printed in the United States of America

18 17 16 15 14 13 12 11 10 09   1 2 3 4 5

ISBN-13: 978-0-226-72082-1 (cloth)
ISBN-10: 0-226-72082-9 (cloth)

Endpapers: Plan Showing Works at Lake Thirlmere (front), Plan of Line of Aqueduct from Thirlmere to Manchester (back). John James Harwood, *History and Description of the Thirlmere Water Scheme* (Manchester: Henry Blacklock, 1895), pp. 30, 78.

Library of Congress Cataloging-in-Publication Data

Ritvo, Harriet, 1946–
The dawn of green : Manchester, Thirlmere, and modern environmentalism / Harriet Ritvo.
p. cm.
Includes bibliographical references and index.
ISBN-13: 978-0-226-72082-1 (cloth : alk. paper)
ISBN-10: 0-226-72082-9 (cloth : alk. paper)
1. Water-supply—England—Manchester—History. 2. Thirlmere, Lake (England)—History. 3. Thirlmere, Lake (England)—Water rights—History. 4. Environmentalism—England, North West—History—19th century.
I. Title.
TD264.M3R58 2009
333.91009427'3—dc22

2008051313

⊗ The paper used in this publication meets the minimum requirements of the American National Standard for Information Sciences—Permanence of Paper for Printed Library Materials, ANSI Z39.48-1992.

For my old friends
Andrew, Jennifer, Henry, George,
and Ellie Warren

"I've seen gardens, compared with which this would be a wilderness."

The Red Queen in Lewis Carroll,
Through the Looking Glass

Contents

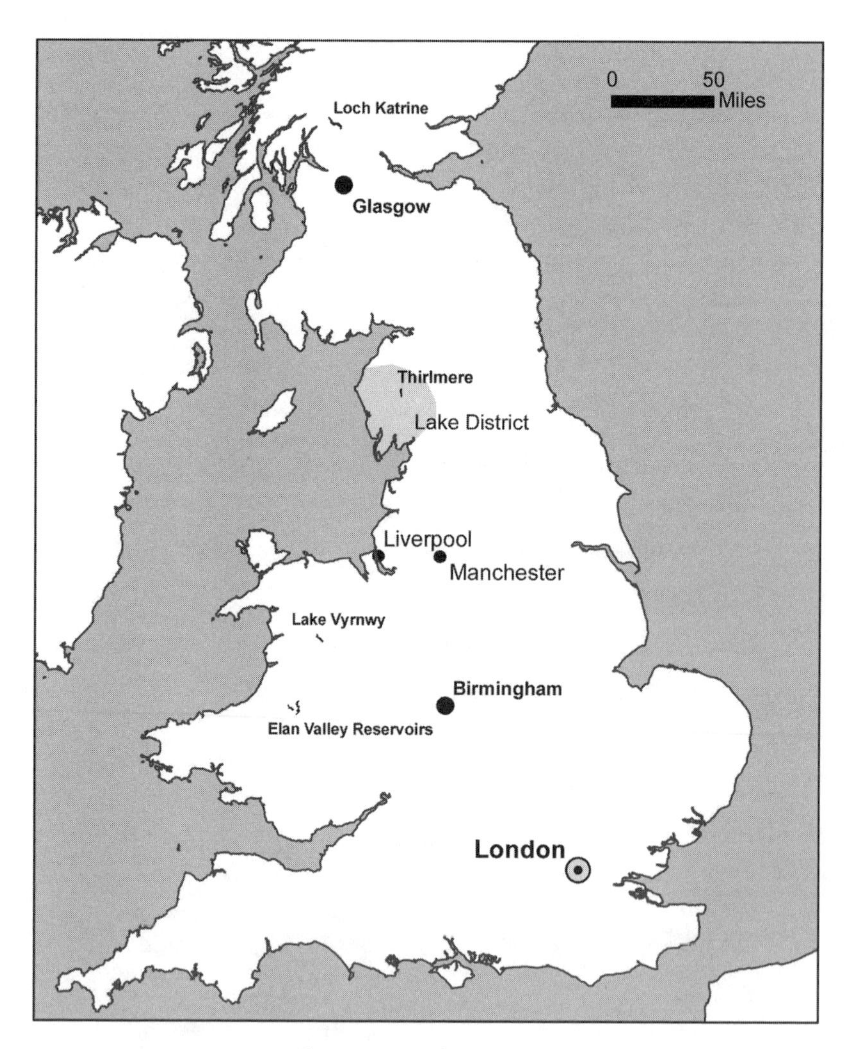

Some cities and reservoirs.

Introduction

Almost every visitor to the Lake District passes Thirlmere, because it lies beside the A591—the main road connecting Kendal and Keswick, the area's principal towns—but few pay it much attention. It looks pleasant but unremarkable, and the water is intermittently obscured by trees. Postcards of it are hard to find, even at the King's Head Inn, situated across the road from the lake, as it was before Thirlmere became a reservoir. A hundred and fifty years ago, Thirlmere looked very different—and much more typical of the Lake District. Indeed, it then presented one of the harshest and most dramatic of Cumbrian scenes, instead of one of the blandest.

While the lake is oddly unobtrusive, the history of its transformation is hardly secret. It can be read on the commemorative plaques erected at the northern end of the lake, beside the road that crosses the massive dam that converted Thirlmere into a reservoir. It can be read, in somewhat greater detail, on the signs posted by the small parking lots and marked trailheads that form a relatively recent (and newly inviting) feature of Thirlmere's public face. And it is routinely mentioned in histories of nineteenth-century engineering, of public health, of landscape protection, and of municipal government. But as the lake itself lacks prominence among the competing attractions of the Lake District, so its story has constituted a relatively minor episode in the annals of Victorian progress.

Large public works were far from unusual in the nineteenth century, and they inevitably presented significant technical and administrative challenges. The construction of railway lines and gas mains, like that of dams and aqueducts, required the permanent transformation of some landscapes and the temporary disruption of others; all such projects involved the acquisition of large amounts of property from reluctant or unwilling owners and the extraction of financial resources from ratepayers or stockholders. Even as a water project, Thirlmere was far from unique. Within several decades of its conversion into a reservoir for Manchester, Glasgow, Birmingham, and Liverpool had undertaken similarly ambitious projects to collect rural water and transport it to urban homes and factories. These other reservoirs are now more conspicuous—their history less modestly acknowledged, their commemorative plaques larger and more prominent.

The relative celebrity in our time of these reservoirs reverses the Victorian situation. Although the water projects of rival cities were admired and reported on, the spotlight of public attention shone more brightly on what was referred to as Manchester's Thirlmere Scheme. This attention was far from uniformly positive. On the contrary, what propelled Thirlmere into the spotlight in the mid-1870s and kept it there intermittently for the next quarter century was the controversy that erupted as soon as intimations of Manchester's plan leaked out. This controversy was different in both degree and kind from the resistance routinely provoked by large public works projects, which came from landowners, ratepayers, and others whose property or other economic interests might be damaged. The Thirlmere controversy engaged a much wider constituency. Lovers of the countryside from throughout Great Britain, and indeed, from throughout the empire and former empire, claimed that they would experience loss as a result of the transformation, even though they had only the most insubstantial claim to preexisting possession.

The Thirlmere Scheme thus offered an early occasion for the expression of almost the full range of modern opinions about large-scale environmental engineering, whether in the ostensible service of public welfare or of private industry. It is not surprising that such a confrontation would occur in the Lake District of northwestern England, or that the threatened encroachment would come from the city of Manchester. Of all the rural regions of England, the Lake District was the most cherished for its natural beauty and its cultural associations. Great Britain was at the forefront of the set of transformations that produced the world of modern industrial capitalism, and within Britain Manchester epitomized this irresistible but troubling process. When Manchester bought Thirlmere in order to convert it from lake to reservoir, therefore, the Victorian icon of nature confronted the Victorian icon of progress. The first two chapters of this book will evoke these icons; the succeeding ones will provide a detailed narrative of the Thirlmere Scheme itself, from its inception and realization through its invocation in the debate about Haweswater, the second Cumbrian lake to be taken by Manchester (in the 1920s), to its nationalization and privatization late in the twentieth century. The conflicts that began in the nineteenth century continue to reverberate at Thirlmere, where, although the main battle was won and lost before 1900, some side skirmishes have yet to be decided. In addition, such disputes have been replayed elsewhere in the intervening century, as conservation and other environmental concerns have gained prominence throughout the world. The fight for Thirlmere was in many ways quintessentially Victorian, engaging particular issues of municipal governance, national economy, and shared English heritage. It has also turned out to be a kind of template for subsequent environmental struggles.

The outline of the story has become so familiar as to seem predictable. Once the Manchester Corporation (the city's governing body) set its sights on Thirlmere, the lake's fate was sealed. It was duly purchased and dammed, its level raised by as much as fifty feet, and its waters piped one hundred miles southeast to the cisterns of Manchester. But the road to this conclusion was much rockier than the developers had anticipated. Local people, with national and international support, organized a campaign of resistance. From a narrowly pragmatic perspective, their crusade was quixotic. Although their voices carried, they were not supported by economic and political clout equivalent to (or even in the same league as) that possessed by the city. But they did manage to cause expense, inconvenience, discomfort, and sometimes personal embarrassment to their powerful opponents. Still more important, the publicity that their efforts attracted both revealed and encouraged a strong vein of public sympathy.

The mere fact of controversy—of the expression of alternative perspectives—does not constitute the major significance of this case, either for the Victorians or for us. What made the Thirlmere Scheme especially noteworthy in its own time, and especially predictive of the shape of future conflicts, was the conspicuous involvement of individuals and interests unconnected with property in the narrowest sense. Thirlmere lay close to the center of the Lake District, which had for a century occupied a preeminent position in the pantheon of English natural beauty, even before its sacred status was consolidated by the poetry of William Wordsworth and his fellow Lake poets. Further, by the middle of the Victorian period, many writers, politicians, and others with ready access to the press had become summer residents of the Lake District—in large part, ironically, because of the construction of a railroad that Wordsworth had opposed a generation earlier. And perhaps most important, the Thirlmere Scheme was broached at a time when the notion of public ownership of landscape was being expanded and consolidated, rendering it both newly potent and newly vague. In tandem with organized attempts to protect physical access to private property, via rights of way or public footpaths, came assertions of a new kind of spectatorial right or lien on the land. It was asserted that the citizenry as a whole—the nation, that is to say—had a vested interest in preserving the traditional appearance of certain rural landscapes. Similar claims underlie contemporary arguments for conservation and preservation.

The nebulous new sense of ownership invoked by such assertions—a sense that citizens of a nation (or, still more expansively and vaguely, members of a supranational cultural community) should have some say in the disposition of significant landscapes even if they held no formal title to the property in question—was both radical and polarizing. Many eminent and eloquent personages entered the lists on each side, most lining up rather predictably. The big guns of the Church of England in Cumbria, for example, were against the Scheme, while those in Manchester favored it. The debate was carried on in a variety of media and fora. It was punctuated by a series of public and semipublic confrontations, of which the most conspicuous and dramatic were the parliamentary hearings, with their parade of personalities and passions. Politics, aesthetics, technology, and science mingled in arguments about the costs and benefits of the Thirlmere Scheme—before it was implemented, after its completion, and still, retrospectively, more than a century later.

Of course, advocates of the Thirlmere Scheme did not argue that they wished to deface an unspoiled valley or to destroy the integrity of a uniquely beautiful and significant region. Instead, they stressed progress and prosperity. Manchester needed more water because it had grown with unforeseen rapidity; it was to be hoped that such growth would continue, and not only for the benefit of the industrialists who reaped its profits and sang its praises most conspicuously. The whole nation had a stake, since the British economy depended heavily on the manufacturing districts of the north. But, as its spokesmen tirelessly repeated, the Manchester Corporation was also acting on behalf of its entire local constituency, ordinary citizens as much as captains of industry. Proponents therefore presented the search for water, in terms that combined populism and paternalism, as a quest to ensure full employment and modern sanitation for Manchester's working classes.

They even challenged their critics on aesthetic grounds, asserting that, far from impairing the Cumbrian landscape, the reservoir would enhance its natural beauty. The carriage road to be built along with the proposed waterworks would make Thirlmere more accessible, so that the best views of the lake, previously restricted to intrepid pedestrians, would become available to all. At the same time, Manchester's plan would paradoxically also preserve Thirlmere from the depredations of tourism and ordinary commerce, which were already evident in other parts of the Lake District.

It may sound as though these two positions, although opposed, were reconcilable. But recognition that the opposing position had some merit was not really the issue. Only the most blinkered of industrialists and engineers refused to acknowledge that Thirlmere, and the Lake District more generally, embodied and represented values that could not be completely assessed in utilitarian terms. Similarly, only the most intransigent of the lake's defenders regarded Manchester's desire for more water as indefensible. Instead, the issue was relative: of two acknowledged goods, in which should have priority? From the perspective of the Thirlmere Defence Association, there was no question that the preservation of the Lake District was more important than supplying Manchester with the best and cheapest water. From the perspective of the Manchester Corporation, the concrete physical and financial requirements of its citizens and factories easily trumped the more nebulous concerns of remoter constituencies.

A century's hindsight only makes the Thirlmere controversy appear more complicated. The assessment of a policy or set of actions must depend to some extent on the range of available options. In 1878 the most compelling alternative to the Thirlmere Scheme was the Thirlmere non-Scheme—the preservation of the status quo. That option, of course, no longer exists. Possible alternatives now are represented by the other large Cumbrian lakes, which exemplify various histories of exploitation and development. Next to Thirlmere, some seem to have suffered at least equal violence, and perhaps for less good causes. Around the world, dams—now built on a scale far beyond the imaginations of Victorian engineers and officials—remain among the most environmentally controversial of development projects. Expanding human population, increasing individual expectations, and economies based on constant growth make it unlikely that the pressures that triggered the initial fight for Thirlmere will become less intense any time soon.

I

The Unspoiled Lake

In 1875 local residents began to sense that property around Thirlmere was attracting unusual attention. First there was unobtrusive surveillance, then cryptic purchases of land. The members of the Manchester Waterworks Committee, who had identified the long narrow lake as a likely site for the city's urgently needed new reservoir, hoped that low-key tactics would discourage suspicion and keep asking prices low. Those hopes were doomed to disappointment. The deliberations that had led them to Thirlmere were not completely secret, and simple prudence required that they examine the potential site of such a significant investment before committing to it. It was difficult to keep such hands-on explorations inconspicuous.

One expedition began tamely enough at Ullswater, a neighboring lake that was also initially under consideration. Having finished its business there by early afternoon, the surveying party decided to walk to Thirlmere over Helvellyn, one of the highest of the Lake District hills. This route proved more challenging than anticipated. The two oldest members of the group, one of whom "could not walk very well" even in ordinary circumstances, eventually began to flag and had to take turns riding the single pack horse, which meant that their fitter companions had to hike with the baggage. At one point, the horse sank to its knees in a peat bog, from which it was extricated only with difficulty, and two of the walkers injured themselves seriously descending a hillside in the dark. They did not reach Thirlmere until ten at night, when they abandoned themselves to the mercies of a local public house. There they waited several hours longer for a carriage to take them to their hotel in Keswick, more than ten miles away.[1]

On another occasion two dignified members of the Waterworks Committee wished to observe the lakeshore from the grounds of the leading local landowner, who had explicitly denied them access and "threatened severe measures if he found any of them ever hanging about his property." They therefore approached the forbidden vantage point by way of the fields, rather than from the road, creeping past the manor house on hands and knees, concealed in soaking vegetation. After completing their reconnaissance, they had to endure the long ride back to Keswick in wet clothes; both became ill and "were laid up for some days."[2] Merely alarming was the experience of one of their colleagues, who ventured onto Helvellyn for an aerial view of Thirlmere one sunny winter afternoon, after completing a purchase of land near the lake, and found himself climbing down through "blinding snow."[3]

In their choice of Thirlmere as the prospective site of a reservoir, and in their first tentative ventures into the Lake District, the members of the Manchester City Council were guided by Alderman John Grave. Born in Cockermouth, where his father was a saddler, he made his fortune manufacturing cotton and paper in Manchester.[4] He had maintained his connection with his native Cumberland and owned a house near Keswick, from which base he would later be accused of "carefully nursing the whole district . . . , buying here and there."[5] But before any buying could be done, it was necessary to learn more about individual holdings, whether large estates or modest farms, than could be gleaned by mere visual inspection. Once his fellow Cumbrians got wind of Grave's plan, however, neither his local roots nor his "very speculative turn of mind" would help him to elicit sufficient (or sufficiently accurate) information.[6] Any identifiable representative of the Manchester Corporation would be stonewalled by canny property owners mindful of future negotiations. So three local ringers received discreet appointments as "special commissioners." All pursued occupations that provided natural cover for peripatetic inquiry: one was a cattle dealer, one a building contractor, and the third (best of all) was

Henry Irwin Jenkinson, a champion fell walker and the eponymous author of *Jenkinson's Practical Guide to the English Lake District.*[7] Since, at least according to an uncharitable retrospective view, his research subjects hoped that he would bring "the innocent and unsuspecting tourist into the locality, for the purpose of being skilfully fleeced, information was freely and readily imparted."[8]

As it turned out, however, financial concerns were not uppermost in the minds of most local landowners when they became aware of Manchester's interest in Thirlmere. And reticent though they might have been about the details of their property, they were more than forthcoming with their opinions of the prospective appropriation. It pleased them not at all. The determined opposition of Thomas Leathes Stanger Leathes, whose property included not only Dalehead Hall, the most imposing home on the lakeshore, but also the lake itself, turned out to be a straw in the wind. Less prominent residents, already nerved to resist a rumored extension of the Lake District railway service into their neighborhood, shared his reluctance to see Thirlmere and its surrounding valley transformed to meet remote urban needs.[9] Proclaiming their allegiance to a landscape conceived both romantically and traditionally, they organized to oppose the forces of progress and materialism, industry and economics, of which Manchester was both the representative and the symbol. The Manchester City Council members and their agents were surprised by the energy of this resistance and by its far-reaching appeal. Their Cumbrian antagonists quickly began to receive moral and material support from throughout the nation, and indeed the Anglophone world. A confrontation that the projectors had expected to play out in familiar commercial terms slipped uncontrollably into the realm of ideology and myth, since if Manchester was the icon of the Victorian future, the Lake District was the icon of nature, poetry, and heritage.

The Lake Itself

Before its emergence as the object of Mancunian desire, Thirlmere had played a relatively modest part in the history of Lake District appreciation. This history stretched back for nearly two centuries. Tourists had begun to visit the remote hills of Cumberland and Westmorland at the end of the seventeenth century, although at first not all were pleased with what they encountered. Daniel Defoe, for example, found the Cumbrian landscape "eminent only for being the wildest, most barren and frightful of any that I have passed over in England, or even in Wales itself."[10] By the

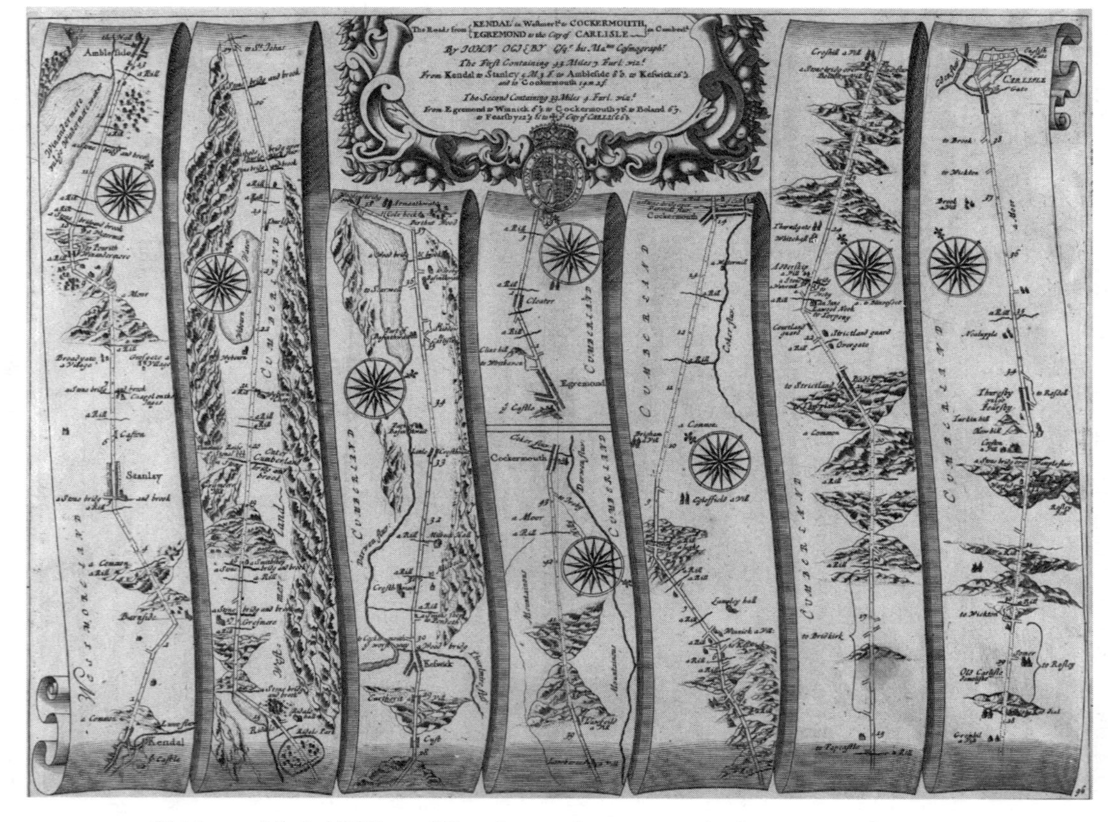

Thirlmere, labeled "Wiburn Water," seemed very remote in the seventeenth century.

middle of the eighteenth century, however, aided by the accounts of more responsive viewers like the poet Thomas Gray, travelers were learning to love the dramatic valleys and rugged hilltops.[11] In addition, the Lake District had become somewhat more accessible as a result of the construction of new roads. Particularly effective in opening up the heart of the area was a turnpike trust established in 1761 to improve the highway between Kendal and Keswick, the most important towns.[12] The northwestern portion of this road, which ran from Windermere (by far the largest of the lakes) to Keswick, was later said to be "the trunk of the District . . . it traverses through twenty-one miles of, perhaps, the most beautiful highway-scenery in England."[13]

By the time that George Culley and John Bailey made their agricultural survey of the lake counties at the beginning of the nineteenth century, they parenthetically noted that the area was "ornamented with many beautiful and extensive lakes; which, with their pleasing accompaniments have of late years made the tour of the lakes a fashionable amusement."[14] Along with tourists came the infrastructure of tourism: inns, carriages, guides, and guidebooks.[15] Aficionados of the picturesque had even begun to build houses so that they could appreciate the scenery on a more protracted basis.[16] As aesthetic sensibilities grew less restrained, the charms of Cumbria became still more compelling. It was because "the scenery of Westmoreland and Cumberland cannot be exceeded in varied grandeur and beauty" that a travel book published in 1821 could claim that "England . . . can vie with any country in Europe in wild and romantic scenery" and, still more strongly, that "in all that constitutes the perfection of romantic landscape, England is without a rival."[17] That is, as a slightly earlier traveler had put it, "Nature" could be judiciously admired in the Lake District without "the eye . . . being either *glutted* by expanse, or DISGUSTED by deformity," hazards that awaited the sensitive English tourist in Switzerland and other foreign destinations.[18] Looking back in 1867, the anonymous reviewer of ten books about the Lake District commented, "It is most curious to contrast the descriptive language with which we are familiar, with the descriptions, expressive rather of repugnance than of admiration, which belong to an earlier time."[19]

In the first decades of the nineteenth century, the physical attractions of the Lake District were enhanced by literary association.[20] William Wordsworth, Robert Southey, and Samuel Taylor Coleridge, who became known as the Lake poets, made the area both their subject and, for varying periods, their home. Wordsworth, like John Grave a native of Cockermouth, had spent most of his schooldays in the isolated vale of Esthwaite, and he returned to the neighborhood of Grasmere, more conveniently situated near the main road, for most of his long adult life.[21] His poetic evocations of the striking Cumbrian landscape and his passionate response to it inspired many readers to make their own pilgrimages. Their advent not only boosted the nascent hospitality industry, but generally enhanced the prosperity of local landowners. As one of them testified much later, during the contention over Manchester's reservoir plans, "Wordsworth has done . . . more than any land improver can do; he has doubled and trebled the value of the property by the associations."[22]

Wordsworth was to remain ambivalent about Lake District tourism and his own role in promoting it. At some times he encouraged it through his writing, while at others he lamented the predictable consequences and even attempted to undo them. He countered the most excitable strain of touristic response implicitly, by example. In 1810 he anonymously published the first version of a guidebook (he took credit for subsequent editions) that offered rather down-to-earth descriptions of the local attractions. He credited his ideal audience, which he imagined to consist of "Persons of taste, and feeling for Landscape, who might be inclined to explore the District of the Lakes with that degree of attention to which its beauty may fairly lay claim," with similarly judicious restraint.[23] But it was difficult to suppress altogether the more extravagant responses of such readers as the essayist Thomas De Quincey, who like many others, was drawn to Cumberland and Westmorland by Wordsworth's poetry. (De Quincey, again like many others, was also drawn to the man himself; by the end of his life, both Wordsworth's house and his person figured routinely in tourist itineraries, along with the settings of his best known poems.) In his enthusiastic reconstruction of Wordsworth's childhood experiences, De Quincey characterized the Lake District as a "domestic Calabria" (probably a mixed compliment, at least with regard to its human population), a "paradise of virgin beauty, . . . even the rare works of man . . . were hoar with the gray tints of an antique picturesque."[24]

Whether attracted to Cumbria's scenery or to its literary refraction, almost all visitors included Thirlmere in their itineraries. Indeed, it was difficult for them to avoid at least glimpsing it, because the carriage road from Kendal and Windermere to Keswick ran along the eastern shore of the lake, shadowed by the massive bulk of Helvellyn. Because Thirlmere lay in a steep, narrow valley—the surrounding hills and cliffs rising abruptly from its rocky shore with little intervening field or meadowland—it presented a less genial view than did such more celebrated lakes as Windermere, Grasmere, Derwent Water, and Ullswater. Thus Thirlmere functioned as a kind of aesthetic litmus test. Responses to it indicated not only the idiosyncrasies of individual sensibilities, but also the increasing relish for the sublime, an increase that seems to have been more gradual with regard to the Lake District than with regard to other places. The harshness of Thirlmere's setting made most eighteenth-century visitors uneasy. Even Thomas Gray found that "the water looks black from its depth (tho, really clear as glass) & from the gloom of the vast crags, that scowl over it."[25] In a guidebook published in 1784, Thomas West tried to make the scene more appealing by recasting it in the less demanding terms of conventional landscape appreciation, with a shore that was "beautified" by scattered rocks and a ring of "verdant mountains" that rose "in the noblest pastoral stile." Even so, the prospect seems to have conveyed at least as much physical threat as visual pleasure: "A thousand huge rocks hang on Helvellyn's brow, which have been once in motion, and are now seemingly prepared to start anew."[26] West raised his eyes to the hills as much in search of avalanches as of inspiration.

His contemporaries tended to be franker about their reactions. James Clarke, a land surveyor, complained that the valley lacked "anything very entertaining"; further, the "tremendous mountains" that flanked it were "laden with large loose stones, which seem ready to drop . . . on the smallest occasion, a sight of sufficient terror to hasten the traveller from a scene of such seemingly impending danger."[27] The artist William Gilpin, who traveled in search of the picturesque, identified the "entrance into Cumberland" (that is, the descent into the valley of Thirlmere from the south) as "a scene very strongly marked with the sublime," and the lake itself as in "every way suited to the ideas of desolation, which surround it."[28] Another visitor referred to its shape as "antlike" and regretted that even the proximity of "lofty Helvellyn" could not prevent the scene from "degenerating" as he passed it on his way south.[29] Even Ann Radcliffe, the celebrated purveyor of the exoticized fictional Gothic, proved less than enthusiastic in the presence of the domestic sublime. She found Thirlmere a "narrow and unadorned lake, having little else than rocky fells," Wythburn at its southern tip a "poor village," and the whole valley a scene of "ruin and privation."[30] The author of a subsequent account, an artist who lived in nearby Ambleside and who was, therefore, more familiar with the area than passing tourists were likely to be, speculated that Radcliffe, like others who shared this general opinion, had "not . . . deviated from the turnpike road"

Victorian picnickers were no longer deterred by Thirlmere's harsh landscape.

and suggested somewhat unpersuasively that she would have felt very differently about the lake had she visited its less accessible western shore.[31]

As time went by, visitors experienced less foreboding and greater delight. Coleridge poetically forecast this reversal in a notebook of 1803, apostrophizing the lake—"O Thirlmere!"—and celebrating the very features that had disturbed earlier visitors: its "naked or ferny Crags—ravines, behaired with Birches . . . dazzling *wet places* of small rock-precipices . . . Cliffs like organ pipes."[32] Within a few decades, this more enthusiastic perception had become widespread among tourists of ordinary sensibility, even verging on a stock response. Describing a trip taken in the late 1820s, the journalist Edward Baines evoked a scene that his eighteenth-century predecessors would have recognized easily—"the dark, narrow lake . . . overhung . . . by crags . . . naked and gloomy"—but he assigned it a positive aesthetic and emotional valence, characterizing the surrounding cliffs as "stupendous" and the whole scene as one of "wild magnificence." He noted with approval that Thirlmere's shore was not prettified by decorative plantations of trees, in contrast with the more cultivated shores of many of its fellow lakes, and with relief that its few stands of fir rather tended to increase its gloom.[33] At about the same time, the author of an article that ranked Thirlmere among the "beauties of England," enthusiastically described it as "the most lonely and desolate of the lakes." It backed up this visual impression with actual perils, both physical (the winding road was "dangerous for travellers," as the slightest deviation might occasion their destruction) and psychological ("the frequent suicides . . . almost make one consider the place dangerous to a person inclined to melancholy").[34]

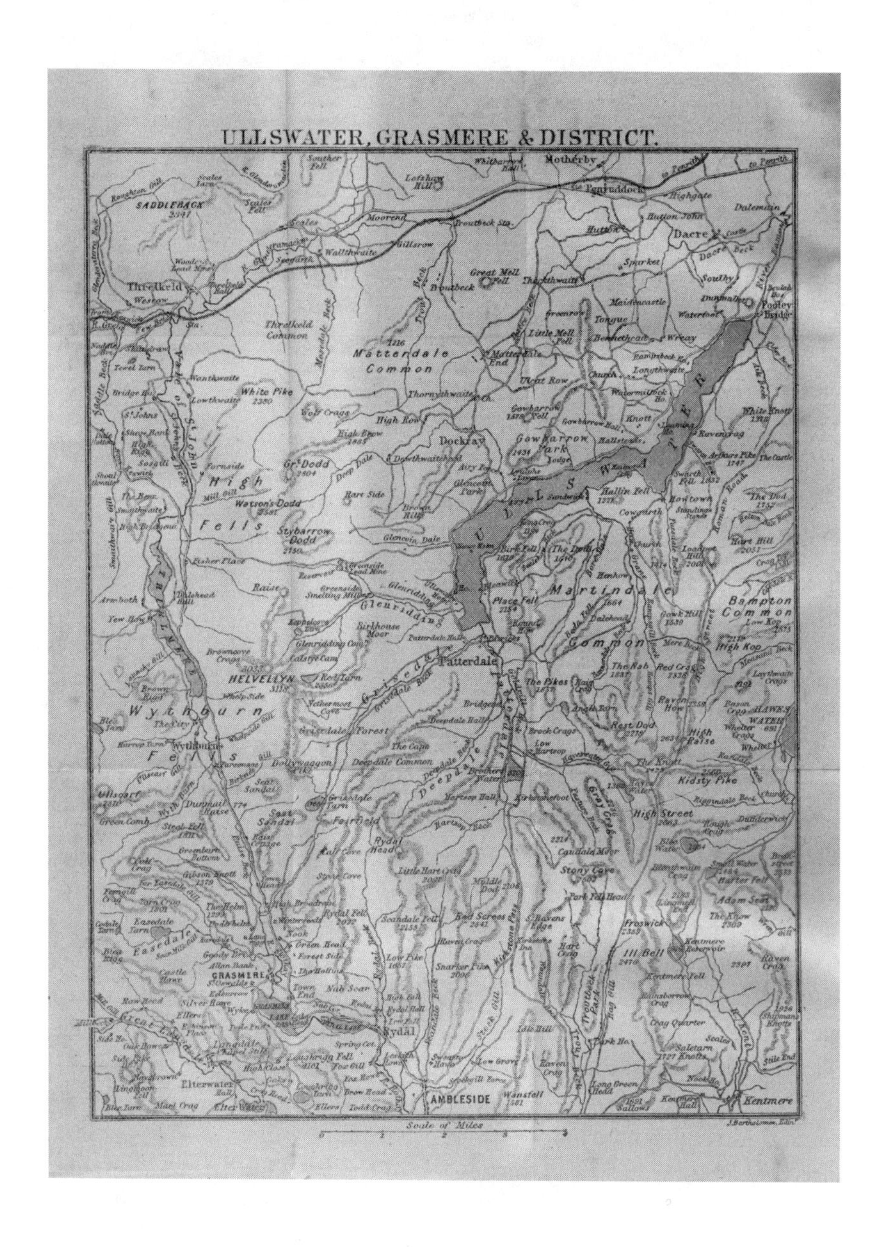

The Rock of Names.

So strongly did subsequent Victorian taste run in this direction, that several decades later Harriet Martineau regretted the disappearance of the dark arboreal masses, complaining that "the only gloom in the scene is from Helvellyn."[35]

Thirlmere also basked, at least to a limited extent, in the grace shed on the Lake District by literary association. The primary stamping grounds of Wordsworth and his friends were near Grasmere and Rydal Water, a few miles to the south along the main road. They walked along Thirlmere's shores and on the surrounding heights, and recorded their pleasure in poems and journals. They enjoyed the hospitality of Dalehead Hall, the owner of which at a later period was to bar inquiring Mancunians. This group—William Wordsworth, his brother John, his sister Dorothy, his wife-to-be Mary Hutchinson, her sister Sara, and Samuel Taylor Coleridge—created (or were understood to have created) what was to become Thirlmere's only literary shrine by carving their initials on a wayside stone, to mark the halfway point between Grasmere and Keswick, where Coleridge was living. They called this "trysting place" the "Rock of Names."[36] In addition to these most renowned visitors, Thirlmere attracted intermittent attention from other literary figures. Much later—after Manchester had taken possession of the lake, but before construction had begun—the ailing Dante Gabriel Rossetti spent a holiday in a nearby farmhouse, writing to friends that "the scenery is the most romantic and beautiful that can be conceived, and the retirement more absolute than I ever met with before."[37] The neighborhood of Thirlmere also figured occasionally, although seldom memorably, as the setting for poetry and fiction. Walter Scott, who had visited the Lake District in 1805, used the adjoining Vale of St. John as the backdrop for his Arthurian narrative poem *The Bridal of Triermain* (1813). The walk that Matthew Arnold sadly retraced in "Resignation" (1849) wound north from the inn at Wythburn. In 1885 Hall Caine set *The Shadow of a Crime,* his novel of seventeenth-century Cumbrian folk life, near Thirlmere, as Mrs. Humphrey Ward did *The Mating of Lydia,* her 1913 novel of modern love and avarice.

Past and Prologue

Despite this consistent modicum of attention and the gradual improvement in its aesthetic rating, Thirlmere never became a primary destination for Lake District visitors, or the cynosure of admiring gazes. A midcentury traveler, who professed to "love it dearly," regretted that it was "the most disregarded [of the lakes] . . . a great injustice."[38] No matter what their

Using Thirlmere water.

assessment of its charms, most travel books simply described the passing view from the road, named the local inns, and identified the starting points for the ascent of Helvellyn. And although Helvellyn ranked among the highest Cumbrian peaks, its most striking profile was not visible from the lake. In any case, according to one artistic observer, it was "one of the least impressive mountains in the Lake District, considering its elevation."[39] More perfunctory or selective guides that noted only the highlights of a Lake District tour often omitted Thirlmere altogether. This relative inattention did not, however, turn out to mean that the lake seemed dispensable in any way. On the contrary, from their beginning, protests against its incipient transformation insisted on the unique and superlative value of what was about to be lost. As the bishop of Carlisle forcefully put it in a letter to the *Times,* "Thirlmere is among the choicest of English lakes. . . . It is as wild as it was centuries ago, and the wooded crags which overhang it are quite unsurpassed in beauty."[40] To the surprise of the Manchester city fathers, this point of view proved very compelling.

If such claims seemed unpersuasive in the light of Thirlmere's conventionally secondary position among its fellow lakes, neither could the nature of Manchester's proposed project account for it. Massive public works projects—involving the construction not only of dams, reservoirs, and pipelines but also of railways, sewers, roads, canals, and large official buildings, all of which were apt to cause enormous disruption—were common features of Victorian urban and rural experience. Such projects normally provoked protests from the people whose interests they directly affected: that is, occupants who would lose their homes and workplaces, either permanently or temporarily, and owners who feared that they would not be sufficiently compensated for the sale or use of their property. But it was rare for such resistance to last long or to deeply engage the sympathies, let alone the pens and purses, of anyone outside this immediate circle.

Certainly, nothing in Manchester's previous experience with the construction of rural waterworks forecast opposition of the kind aroused by the proposed conversion of Thirlmere. This experience had been significant and far from untroubled. In the late 1840s the Manchester Corporation began building what was to become a massive series of reservoirs—the largest system of artificial lakes in Europe when it was completed—in the Peak District, about twenty miles to the east of the city.[41] The Longdendale Scheme, as it came to be known, aroused fierce controversy in its planning stages; its engineer, John Frederic LaTrobe Bateman, and the politicians who had engaged him had to defend the ambitious project vigorously at every turn. But the challenges came from predictable quarters. When the enabling legislation was considered by Parliament in 1847 it was formally opposed only by the Sheffield and Lincolnshire Railway Company, a private concern that had hoped to supply Manchester with water on a for-profit basis, and by a group of mill owners whose factories depended on water from a local river. Ultimately Parliament guaranteed that the mill owners would be provided with compensation water as part of the reservoir system and allowed the railway company to sell water in bulk to the Manchester Corporation but not to compete with the city as a supplier. The legislation also allowed the Corporation to levy compulsory water rates or charges on both occupiers and owners of residential and business premises, thus addressing (if hardly assuaging) a source of discontent ineluctably inspired by any large civic investment. After construction had begun, Bateman confronted a series of technical challenges: the underlying geology of the Longdendale area turned out to be less stable than it looked, and, more generally, many important questions about hydraulic engineering had yet to receive definitive answers in the mid-nineteenth century.[42] These problems consumed time, money, and political capital, but they were all susceptible to concrete pragmatic solutions. (For a fuller discussion of the Longdendale Scheme, see chapter 2.)

At the same time that his elaborate plans were being realized at Longdendale, Bateman was also working on a new water supply for the city of Glasgow. Like every other burgeoning industrial city of the nineteenth century, Glasgow had struggled to provide its citizens with water sufficient in amount and quality for their domestic, commercial, and industrial needs. In 1852, after decades of unhappy experience with private water companies, the city council decided to take matters into its own hands and underwrite a municipal reservoir. It hired Bateman as a consultant in the early planning stages, and followed his advice by choosing Loch Katrine, about thirty miles to the north, as the source of additional water.[43] Bateman was ultimately appointed superintendent of the works and oversaw the many challenges of construction until 1859, when Queen Victoria ceremonially turned the spigot that sent the first reservoir water on its way to Glasgow taps. At a banquet given in his honor, Bateman praised his employers for their "wisdom" and "liberality" in sponsoring "a work which surpasses the greatest of the Nine Famous Aqueducts which fed the City of Rome."[44]

In its general outlines, the Loch Katrine Scheme significantly resembled the plan that Bateman was later to promote for Thirlmere. It involved the damming of a rural lake to raise its water level and the construction of a pipeline to transport the water to urban consumers. Technically, too, the plans were similar. As in the case of Longdendale, the Loch Katrine proposal inspired many objections. The most vigorous resistance came from the private water companies whose inadequacies had inspired Glasgow to look elsewhere. In addition to formulating rival plans, they encouraged other objections to the Scheme. Municipal ratepayers complained about the anticipated expense of the project, both in general and in detail, and about the council's new power to levy compulsory water charges. The chemist Frederick Penny raised fears that the soft Loch Katrine water would leach lead from the pipes. The diversion of the water south to Glasgow, away from its original drainage to the southeast, inspired several distinct concerns. Owners of mills on the River Teith required compensation water. Proprietors of salmon fisheries worried that the diminished flow into the River Forth would affect their catch. And the Lords Commissioners of the Admiralty feared that the new patterns of water flow would ultimately interfere with navigation in the Forth estuary. A few of these objections were powerful enough to be heard at the national level—indeed the naval objections delayed the passage of the enabling legislation, introduced in 1854, until the following year. But, as had been the case with the Longdendale Scheme a few years earlier, each potential obstacle was surmounted pragmatically, with the assistance of hard cash or of hard facts.[45]

Although a retrospective critic might group Loch Katrine with Thirlmere, claiming that both "have been violated, and all the other English and Scottish lakes will be similarly ravaged," there was no hint of such resistance at the time.[46] Some of the reservations about the Loch Katrine Scheme were deeply felt and vociferously expressed, but none of them triggered the kind of broad-based and emotional opposition provoked by the

Thirlmere Scheme or engaged the same kind of intangible values. *Punch,* which would rally enthusiastically to the defense of Thirlmere, even published a doggerel celebrating the completion of the earlier project, albeit with an unflattering allusion to a supposed northern aversion to washing:

> Glasgie chiels, a truth ye'll learn,
> New to many a Scot, I'm thinkin':
> Water, aiblins, ye'll discern
> Was na gien alane for drinkin'.[47]

In particular, Loch Katrine's established status as a celebrated beauty spot never became a problem, even when magnified by patriotic literary associations. Walter Scott had used it as a setting in his novel *Rob Roy* (the hero had been born beside the loch) and in his narrative poem *The Lady of the Lake,* where it is described in high romantic style as "one burnished sheet of living gold" surrounded by "mountains that like giants stand / to sentinel enchanted land" and "crags, knolls, and mounds, confusedly hurled, / the fragments of an earlier world" (canto 1, stanza 14). Scott's works had had at least as much impact on Loch Katrine, and the Trossachs region of which it was a part, as had those of Wordsworth on the Lake District. Indeed, they had virtually created the Trossachs as a destination for English tourists, and this debt was everywhere acknowledged. Yet in this case the intervention of technology seemed only to enhance the region's aesthetic and cultural appeal. Even the steamer that plied the reengineered loch during the Victorian period was called the PS *Rob Roy.*[48] After its conversion, Loch Katrine continued to figure in guidebooks as "this beautiful Perthshire lake," its new status as "the principal fountain source of the Glasgow water supply" complementary rather than oxymoronic.[49]

If the Manchester planners had looked to the Lake District, however, they might have anticipated the trouble that awaited the Thirlmere Scheme. They would have uncovered a history of protest against technological incursions. The first flush of railway development, in the 1830s, did not reach the Lake District, but by the early 1840s planning had begun for a branch of the main line from Lancaster to Carlisle, which would pass through Kendal en route to Windermere. Wordsworth was the most conspicuous objector, contributing first a sonnet and then several long letters to the *Morning Post.*[50] The sonnet famously began, "Is then no nook of English ground secure / from rash assault?" and went on to characterize the proposed railway line as "blight," "ruthless change," and "false utilitarian lure." He was especially worried about the damage that might be done to the Lake District scene by the flood of impecunious tourists, for whom it would become newly accessible. Wordsworth's grandiloquent arguments against the new line inspired little public enthusiasm and had no practical effect. Although his position as Poet Laureate gave him ready access to the press, it did not guarantee a sympathetic reception for what was generally perceived as both moralistic opposition to progress and selfish elitism.[51] In any case, his attempted intervention came too late. Supported by powerful local interests, the Windermere Railway was already in an advanced state of planning by 1844, and it transported its first passengers in 1847. Its effects would have confirmed Wordsworth's fears, had he lived long enough to observe them. In 1852 the author of *Adams's Pocket Guide to the Lake District* noted that the need for such literature had been "more particularly felt since the extension of railway communication enlarged the number of summer excursionists." By 1883 the line brought eight thousand trippers to the town of Windermere, itself a creation of the railway, on a single holiday.[52] But if Wordsworth had definitively lost this particular battle (his last, as it happened), the war turned out to be a different matter. As time went by, his ideas became increasingly attractive, especially when events began to make his worries seem prophetic. In the quarter century that followed the opening of the Windermere line, the isolation of the Lake

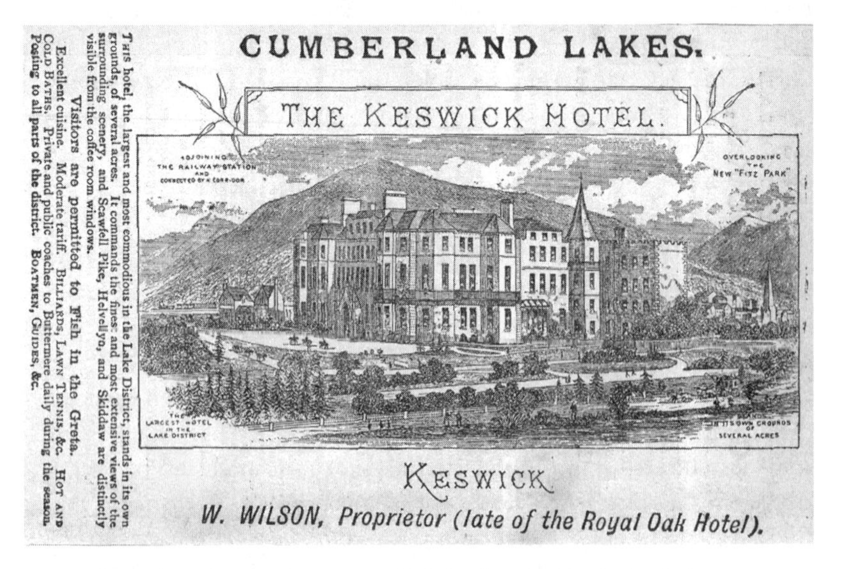

By the late nineteenth century, the Lake District hospitality industry was well developed.

District was further compromised by the construction of three additional railway branches.[53] Like the Windermere line, these stopped short of penetrating the heart of the Lake District, but finally one of Wordsworth's worst fears seemed about to be realized. He had predicted that the Windermere terminus would prove only temporary, and in 1875, according to Robert Somervell, "vague talk was in the air of a possible extension . . . to Ambleside, and perhaps further."[54]

The purpose of the rumored line was to tap the modest mineral deposits of Helvellyn, made newly valuable by the inflation that followed the Franco-Prussian War. The young Somervell, a scion of a prosperous local shoe manufacturing family, "felt, like Elihu in the book of Job, that I must speak out or burst."[55] He accordingly organized a petition to Parliament, then financed a pamphlet that included an impassioned preface by John Ruskin, more or less supportive articles from the *Saturday Review* and the *Daily News,* and a poem from *Punch* ("Were it not well source of such joys to save / Nor wholly yield old Pan a helpless slave / To Puffing Billy?"). At the center was Somervell's own reasoned consideration of the rival claims of economy and scenery, in which he bridged past and future resistance to Lake District development. Beginning with Wordsworth's sonnet of 1844, his essay broached all the themes that were to be elaborated in the struggle against the Thirlmere Scheme.[56]

It is difficult to assess the effectiveness of Somervell's struggle against the railway. Despite some attention from the national press and the support of a few distinguished individuals, Somervell did not receive enough donations even to recoup his investment in the pamphlet. The railway was never built (and indeed no subsequent plan to extend rail service into the central Lake District ever came to fruition), but the project was abandoned because it seemed likely to be unprofitable, not because its backers were daunted (or persuaded) by local opposition.[57] Nevertheless, Somervell's lonely crusade served as a trial run for what was to be a larger and, in some ways, more successful struggle. And it offered the Manchester projectors, had they looked, a clear indication of what was to come.

The Part for the Whole

The members of the Waterworks Committee might also have been better prepared for the resistance to their proposed appropriation of Thirlmere if they had considered the issue in a slightly larger context. For although the lake itself might have seemed unremarkable, the opponents of Manchester's plan did not evaluate Thirlmere in isolation, strictly on the basis

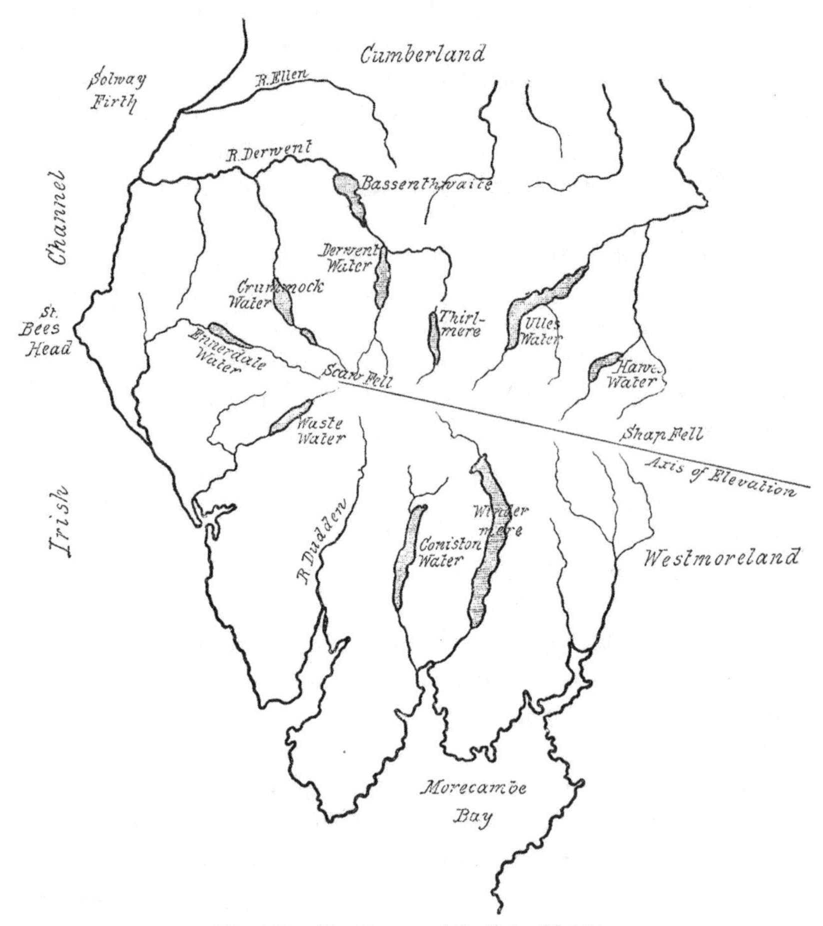

Fig. 163.—Sketch-map of the Lake District.

of its own unique attributes. The seeming extravagance of their claims reflected a sense that the planned transformation of the lake would imperil a much larger landscape. The Cumbrian lakes were conventionally evoked as a group; therefore any one of them could plausibly represent the region as a whole. The slippage from a threat to Thirlmere to a threat to the entire Lake District was easy, indeed, almost inevitable. It occurred repeatedly in the arguments of the lake's defenders. For example, the *Saturday Review* concluded an attack on Manchester's plan for Thirlmere by asserting that "the Lake district must not be spoilt," and the *Builder* dismissed as "rubbish" claims that the Scheme "will not spoil the district or . . . the lake."[58] Despite its high literary and symbolic profile—and the long time required for travel within it—the Lake District was rather small—thirty miles square at a generous estimate. As one vigorous opponent of the Scheme noted, "one can walk across it in almost any direction in a single day."[59] Thirlmere constituted a vital link in a chain that bound together the entire area, not only because of its central geographical location and its conspicuous position beside the major road, but because the preservation of every part was essential to the preservation of the whole. From this perspective, the Lake District could only maintain its rugged beauty by maintaining its territorial integrity; if divided into a set of isolated beauty spots, it would lose its distinctive character and ultimately be destroyed.

The natural history of the Lake District supported this perception of distinctiveness. Because it included the highest points in England (although they were topped by peaks in both Scotland and Wales), the area inevitably attracted the attention of geologists. Its rugged center was described by the Cambridge geologist John Edward Marr as "a rude central dome" composed of Paleozoic rocks, "surrounded by a ring of newer rocks as a rule giving rise to lower ground" (these younger strata were still pretty old, dating from the first part of the Mesozoic era); according to a Victorian contemporary it was "physically and geologically . . . an isolated reproduction of North Wales."[60] The areas of greatest geological interest were, not coincidentally, also those of greatest touristic appeal, and from the beginning scientists suggested that their analyses could complement aesthetic responses. In his 1825 guidebook, Jonathan Otley, whom Marr characterized as "the Father of Lakeland geology," hoped to "direct . . . the tourist through the . . . varied scenery, and at the same time afford . . . information on the structure and component materials of these interesting regions"; he regretted that previously available guidebooks were "wholly devoted to the picturesque appearance of the country."[61] His successors more firmly suggested that technical insight was essential for intelligent appreciation of the dramatic landscapes. As a Geological Survey publication put it in 1876, "even the rudiments of geological knowledge" led the casual visitor "at once to enter into the meaning of the beautiful variety in the scene around him."[62] More sophisticated observers could explain superficial beauties in subterranean terms. Marr associated each of the three types of Paleozoic rock that constituted the central Lake District with a different kind of scenery: the Skiddaw slates and the Silurian rocks supported much tamer hills than the "generally craggy" heights that rose on top of the volcanic Borrowdale series.[63] Thirlmere lay within this most dramatic geological zone.

The unusual topography and geology of the Lake District led to its further differentiation from surrounding territories. Northwestern England was one of the rainiest parts of Great Britain, and inside this damp region the Lake District was particularly wet, with the rugged central area wetter still.[64] Although the climate of the area as a whole was relatively mild, the high elevations of the Cumbrian hills could produce very severe winters—a pentimento of the Pleistocene, when glaciers had formed there and radiated out over lower terrain.[65] These harsher microclimates supported distinctive assemblages of living organisms. When William Hodgson published the first comprehensive flora of Cumberland in 1898, he identified 1,196 species, of which about a hundred belonged to either the "Scottish" or "Highland" geographical region; that is, their Lake District populations were discontinuous with the rest of their species' home ranges. These plants had not been introduced by people. Rather, they were relict populations from the much colder postglacial period that had hung on at high elevations, while lower down they were replaced by plants better adapted to a temperate climate.[66] According to Marr this sturdy flora constituted a "a remarkable assemblage of plants of an alpine character."[67] The faunal assemblage was less idiosyncratic, but the isolation of the Lake District and its lack of intensive development meant that some species, like the otter and the red squirrel, were able to survive there when they had been persecuted or outcompeted elsewhere.[68]

The Construction of Nature

The distinctiveness of its weather, rocks, plants, and animals bolstered the frequently expressed sense that the landscapes of the Lake District were not only unique but uniquely natural. So powerful was this conviction that it could transcend even explicit contradictions. Thus Wordsworth devoted one of the main subdivisions of his *Guide to the Lakes*—entitled "Aspect of the Country as Affected by its Inhabitants"—to a survey of human impact on the Lake District, from the ancient Britons to the early nineteenth century. Although he noted a variety of conspicuously artificial structures, Wordsworth concluded that "the hand of man has acted upon the surface of the inner regions of this mountainous country" in a way that was "subservient to the powers . . . of nature." His conflation of human and natural processes became increasingly thorough as he descended the social ladder. Indeed, the cottages of the humbler inhabitants reminded him "of a production of nature, and may . . . rather be said to have grown than to have been erected; . . . so little is there in them of formality, such is their wildness and beauty."[69] The conventional synecdoche insured that Thirlmere would be similarly appreciated as a profoundly natural landscape, unpeopled and unaltered by human activity. Sometimes observers expressed this conviction obliquely, by describing scenes composed exclusively of natural features. When the young John Ruskin and his cousin Mary Richardson visited the lake in 1830, they noted the height of Helvellyn, as well as "numerous beautiful crags . . . sunbeams . . . little streams" and the "glassy waters of Thirlmere." Human manufacture entered only figuratively, as the "smoothest mirror" to which they compared the surface of the lake.[70] Or the assertion could be explicit, as when Stanley Hughes LeFleming, the proprietor of Rydal Hall, a distinguished residence located several miles south of Thirlmere, petitioned Parliament against the Thirlmere Scheme, claiming that he was "peculiarly interested in the preservation of the Lake District in its natural state, unmarred by modern engineering works."[71]

Thirlmere was sparsely populated and lightly developed, especially considering the proximity of the main road and in comparison with such better known lakes as Windermere, Ullswater, and Derwent Water. These qualities appealed both to the Manchester planners, who hoped that they would translate into easier bargaining and lower prices, and to admirers of romantic scenery. But a sparsely populated area was not the same as an empty one. In the 1870s, and for centuries before, almost every level of rural society and every kind of rural activity was represented in the neighborhood of Thirlmere. A few families, all of whom were to play important parts in the acquisition story, controlled most of the lakeside property. The Leathes family, which owned the lake itself (the water only) in addition to its terrestrial acres, had inhabited Dalehead Hall since the reign of Queen Elizabeth. Although it lacked the grandeur of an aristocratic country seat, it was a "beautiful old house" with an impressive literary pedigree; earlier in the century, Wordsworth and Southey had been frequent visitors.[72] Across the lake from Dalehead Hall was Armboth House, which had been the home of the Jackson family for about the same stretch of time. It was less imposing than Dalehead Hall—really just a substantial farmhouse—but, at least according to a legend frequently rehearsed in Victorian guidebooks, it could claim a supernatural distinction: it was supposedly haunted by a drowned bride, who resurfaced from time to time with a large dog, swimming shoreward to partake of her "ghostly wedding feast."[73]

The head of the Leathes family was the lord of the manor of Legburthwaite, on the northeastern side of the lake, and the head of the Vane family, whose seat was elsewhere in Cumberland, was the lord of the adjoining manor of Wythburn. By the middle of the nineteenth century they lorded it over about two hundred residents of the Vale of Thirlmere, most of whom

Dalehead Hall, Thirlmere.

lived on the less hilly areas near the road, either south as it headed toward Dunmail Raise (the boundary between Cumberland and Westmorland) or northwest toward the opening of the Vale of St. John.[74] In addition to isolated cottages, there were a few small settlements. At the southern end of Thirlmere was Wythburn, site of what was at one time (mistakenly) celebrated as the smallest church in England (the only building in the vicinity of Thirlmere that Nikolaus Pevsner thought worth mentioning in his architectural guide to the region).[75] In recognition of its local preeminence it was also referred to as "the City," although whether with hubris or as a joke was not clear, at least to the authors of guidebooks; the tiny church, analogously, was sometimes called "the Cathedral."[76] Travelers also disagreed about Wythburn's other characteristics. It looked "cold and comfortless" to Adam Walker in 1791 and still seemed "a miserable hamlet with nothing curious about it" when Eliza Linn Linton visited seventy years later. But just before the Thirlmere dam was completed, Samuel Barber called it "a marvelously picturesque spot," noting that "the old world character of the buildings greatly enhances its picturesque charm."[77] Its two inns, the Nag's Head and the Cherry Tree, were generally noted with approval, however. Another well-appreciated inn was the King's Head at the village of Thirlspot or Legburthwaite, about halfway up the lake's eastern shore, across the main road from the grounds of Dalehead Hall.

Although they had been established long before the age of recreational travel, the inns (and the members of their staffs) throve on the increasing numbers of Victorian tourists, especially those energetic enough to wish to climb Helvellyn but not bold enough to attempt the more challenging ascent from its eastern slope. By the middle of the nineteenth century the population included clergymen, a schoolmaster, several tradespeople, and such artisans as carpenters, blacksmiths, stonemasons, and shoemakers, but most local inhabitants continued to live off the land in one way or another. Farming was the most common pursuit. The relatively small amount of flat land at the bottom of the valley was planted with grain (mostly oats) and hardy vegetables, although the growing season was short and the soil could be described as "generally barren."[78] The real agricultural wealth of the district lay in the still less fertile hills and fells, where grazed enormous flocks of the local Herdwick breed of sheep. These small sturdy animals were well adapted to the demanding environment and, in addition to their physical hardiness, were considered to inherit instinctive knowledge of their native terrain from their mothers. As most of the local farmers were tenants, renting the fields they tilled from one of the large landlords, so they also leased grazing rights or stints on the heights according to a complex traditional system.[79] After farming, the most significant local extractive industry was mining. Small-scale workings, mostly for lead but also for iron and copper, had existed on the slopes at both sides of the lake from the

The vale of Thirlmere supported a few farmers.

Champion Hardwick Ram, "Nero," 1870.

beginning of the nineteenth century; by midcentury there was also a commercial-scale lead mining operation on the Thirlmere face of Helvellyn.[80]

The landscape surrounding Thirlmere bore traces of all of these people and pursuits. Most conspicuous were the buildings, those gathered in the hamlets as well as isolated dwellings, sheds, and barns. The valley was crisscrossed by paths and trails, but only the main road was passable by horse-drawn carriages. The paths primarily served the purposes of residents, connecting farmsteads with the hamlets and the main road, or with each other and the upland sheep walks. But tourists also made use of them, if they heeded the advice offered by the more judicious and elaborate guidebooks to view Thirlmere and Helvellyn from the rougher track along the western shore, or if they scaled Helvellyn, or if they undertook one of the more extended pedestrian tours of the Lake District, for which, by the middle of the nineteenth century, there was already a great deal of published advice. Field boundaries were marked by laboriously assembled dry stone walls, and evidence of repeated patterns of sowing and plowing remained imprinted in the earth itself. Active and abandoned mines dotted the hillsides, often to the regret of unwary walkers. And the wooden bridge—variously characterized as Celtic and primitive, as well as "quaint" and "highly picturesque"—that spanned the lake at its wasp waist distinguished it from "all the other lakes in Cumberland and Westmorland."[81]

These were only the most recent and conspicuous evidences of the extent to which human activity had shaped a landscape that still struck observers as wild and forbidding. Such relatively superficial modifications were part of a much longer and deeper history, one that Thirlmere shared with the rest of the central Lake District. But as the celebration of nature tended to obscure living inhabitants and their impact on the landscape, it similarly tended to overlook their vanished predecessors. By the time that the railroads and waterworks began to threaten its natural isolation, people had been exploiting the Cumbrian hills for ten thousand years, since the retreat of the latest Pleistocene glaciation. The ice had hung on longer there than in less elevated regions, and it had visibly marked the land, scouring valleys, depositing stones, creating lakes, and rearranging watersheds. During the Mesolithic period (the time between the disappearance of the glaciers and about 4000 BC), human impact on the Lake District was slight. Contemporary hunter-gatherers were in the habit of burning vegetation, and on the coastal plain to the southwest they did so with sufficient frequency and vigor to alter the composition of the flora, but the uplands retained their tree cover of hazel, oak, pine, birch, and elm (different species predominating at different elevations), which extended to within two thousand feet of even the highest peaks.[82] In the succeeding Neolithic period, people began to settle, even on the less promising uplands, and by the second millennium BC (the Bronze Age) pollen analysis suggests that there had been significant forest clearance, for both pastoral

The "Celtic" bridge.

(grass) and arable (cereal) farming, as well as to provide fuel.[83] Sediments preserved at the bottom of Thirlmere indicate cycles of clearance and regrowth of the surrounding woods as early as 1000 BC.[84] The results of pollen analysis were not, of course, available in the nineteenth century, but these early agriculturists also left less subtle mementos of their presence. Thirlmere is surrounded by prehistoric monuments, ranging from modest cairns and boulders, to the remains of early British hill forts, to the large stone circle at Castlerigg, spectacularly situated within a bowl of hills.[85] Not far south of Thirlmere lay an important center of Neolithic industry: axe heads manufactured from Langdale stone have been unearthed throughout Britain, indicating that this remote area was already embedded in a far-flung trading network.[86]

Despite its proximity to Hadrian's Wall, the Romans did not show much interest in the Lake District. Some of the high Cumbrian roads reflect Roman engineering, but in general, British patterns of settlement and land use persisted, with cultivation and forest clearance fluctuating up and down the hills as the climate became milder or more severe. The English and Danish invaders who succeeded the Romans shared their disinclination for mountain living, and so the central Lake District remained a peripheral Celtic preserve.[87] But the next wave of invaders—the Norse, who arrived in the tenth century—included people well accustomed to scratching an agricultural living from barren uplands. They settled the high valleys with a thoroughness to which the names of local people and places offer continuing testimony—and apparently without too much violence, since much of this relatively undesirable land had remained untenanted.[88] They cleared woods in order to make farmsteads, and their cattle and sheep grazed the hills. Over time, the more efficient and less choosy sheep became the dominant ungulate presence; goats and pigs were ultimately barred in many areas as too destructive.[89] Thus people and livestock together consolidated the dramatic landscape of bare fells and long views that seemed so natural to Wordsworth and his friends half a millennium later.

This stark romantic setting also included some trees, but the woods that Wordsworth admired were very different from those that the Norsemen had felled, let alone those burned by much earlier farmers and hunter-gatherers. As William Gilpin had reflected a generation before Wordsworth, "even the wild features of nature suffer continual change . . . above all, from the growth, or destruction of timber." He attributed the "devastation" of the timber of the northern Lake District to a proximate political cause. After the Jacobite uprising of 1715, the Earl of Derwentwater's estates had been forfeited to the Crown, and their trees had been clear-cut to increase the endowment of the naval hospital at Greenwich.[90] In fact, however, he was describing only the most recent manifestations of a process with much older roots and more varied motivations. Woods had been decaying and falling since people made their first appearance. The deforestation begun by the earliest settlers had proceeded without any intentional check throughout the medieval period, although it was occasionally interrupted or even reversed by forces beyond human control: the vicissitudes of climate (unusually cold or wet periods made marginal lands still harder to cultivate) or of epidemiology (reduced population led farmers to abandon less rewarding terrain). When small-scale local industries became established in the Lake District, as they did in many rural areas where farming was difficult, trees were as likely to be cut to fuel furnaces as to create new fields.

The industrial consumption of trees had very different implications for the landscape, however. Agricultural lumbering was, on the whole, unidirectional; once a field was cleared, it tended to stay cleared, its timber a windfall bonus rather than a renewable crop. Furnaces, however, required a constant supply of wood or charcoal. When, in the sixteenth century, shortages began to loom, it was the smelters, tanners, and other industrialists who attempted to safeguard this essential resource, by protecting existing trees, planting new ones, and promoting sustainable methods of wood harvesting, such as coppicing.[91] The fact that highly industrialized rural counties like Cumberland tended to be much more heavily forested than intensively farmed areas testified to the success of their strategies.[92] Of course, in a sense, these woodlands had been preserved more in principle than in practice. Since their purpose was to provide fuel, they lacked the permanence as well as the biological diversity of their predecessors; stands of trees were repeatedly cut down and monoculturally replanted, with species dictated by botanical fashion rather than by indigenous natural history. In the century before the Thirlmere Scheme was mooted, most descriptions of the lake included some notice of the nearby woods. Their lack of consistency is striking. In 1773 a traveler noted only "scattered trees" around the margin of the lake, and two decades later it seemed "beautiful though unadorned with trees." Nevertheless, Victorian observers looked back nostalgically to a time in the youth of people who were old when Wordsworth was a boy, when "the squirrel could go from Wythburn to Keswick (six miles) on the tops of trees."[93] This recollection might have been accurate—that is, there may well have been a time when squirrels could complete that journey arboreally—but other squirrels attempting the same route, earlier as well as later, would have had to pick their way along the ground.

Thus the woods and fields that evidenced ten thousand years of human occupation of the Lake District became components of the pristine landscape. But if economic activities involving animals and plants were frequently conflated with nature (especially by people who did not make their living from the soil), mining for copper and lead was seldom considered in the same autochthonic light; both its inorganic products and its associated machinery allied it irrevocably with technology. If the emphasis was on the miners rather than on their work, however, even this industrial practice could be naturalized—that is, integrated into the picturesque antiquarian understanding of Cumbrian folk life. There had been mines in the Lake District for thousands of years, but, somewhat oddly, the excavations most frequently transfigured in this way were staffed in the mid-sixteenth century by a group of skilled workers from the neighborhood of Augsburg, in southern Germany, and patented by Elizabeth I as the Company of Mines Royal. Their smelting operation was located in Keswick, and most of their mines were nearby, including one in the Vale of St. John and one at Wythburn. These Thirlmere mines were among the less lucrative of those exploited by the company, which, by the time it closed in 1651, had extracted metal valued at almost seventy thousand pounds, an impressive sum for the time.[94]

Many native Cumbrians must have benefited from this prosperity, especially when the Germans abandoned their initial isolation and began to mingle with the indigenous community, introducing, as one local chronicler put it, "a curious foreign element" into the preexisting demographic mix.[95] The miners left other traces of their activities as well, besides a few exotic surnames. The most frequently adduced reason for the closure of the company was the destructiveness of Oliver Cromwell's troops, but it is also possible that its smelteries exhausted the local supply of fuel. In any case, they consumed a lot of trees; it took some time for the woods to regenerate on the slopes they had laid bare and for the grass to cover the spoil heaps near their pits.[96] Nevertheless, the German miners did not live in memory as protoindustrial despoilers, but rather as quaint contributors to local color. Three or four centuries was long enough for retrospective naturalization. Even Canon H. D. Rawnsley, who was to emerge as one of the most determined and voluble defenders of Thirlmere's status quo, had a soft spot for them. He devoted a whole chapter in a volume nostalgically entitled *Past and Present at the English Lakes* to this Elizabethan enterprise. Acknowledging that "it is true that the German miners denuded the

Lake District of trees," he nevertheless portrayed them sympathetically, perhaps because they made "riches pour into our dales" or perhaps because they "re-opened what probably were old mines."[97] Thus he opened a perspective in which the miners' state-of-the-art operation, which both introduced new techniques to the area and embedded it in the emerging commercial networks of Britain and Europe, could be understood as traditional, with roots in a remote and misty past. From Rawnsley's perspective, and that of many of his coadjutors, the Lake District was by definition unspoiled, no matter how it had come to be that way.

This indulgent transmogrification did not extend to the numerous commercial mines and quarries operating in the nineteenth-century Lake District. The lead mines that dotted the slopes of Helvellyn, for example, had never figured among the beauties of Thirlmere, nor were they adduced as evidence of its "natural state."[98] If they were noted in appreciative descriptions of the lake, it was only to be regretted or disparaged. This was one point upon which there was no disagreement between the Manchester planners and those whose interest in the lake was primarily aesthetic. Although they differed sharply in their understanding of what made Thirlmere pure and pristine, they concurred in attributing a high value to these qualities. Thus in his parliamentary testimony Sir Joseph Heron, the town clerk of Manchester, emphasized that the mines would have to be closed in order to maintain the quality of the water, even though some of his contemporaries thought that a little lead actually enhanced the clarity of mountain streams.[99] The only protest against this feature of the Thirlmere Scheme came from the proprietors of the mines, who accurately foresaw that the proposed reservoir would interfere with their operations.[100] In the end, this was the one consequence of Manchester's transformation of the lake that inspired neither complaint nor criticism. "After all," one commentator grudgingly concluded shortly after the reservoir was completed, "it is somewhat of a satisfaction to feel that . . . the mines of Helvellyn are now likely to remain closed for our generation."[101]

2

The Dynamic City

The scouting parties that occasionally came to grief on the slopes above Thirlmere were drawn primarily from the Waterworks Committee of the Manchester City Council. Since this committee handled some of the most important municipal business, its members included some of the most influential men in Manchester. These dignified representatives of Britain's leading commercial city—"the center of the great cotton-manufacturing industry, and one of the most important places in the kingdom," as a contemporary guidebook unsurprisingly put it[1]—did not ordinarily expose themselves to the physical hazards that inevitably attended clandestine surveillance and trespass in remote areas, still less to the concomitant risk of public embarrassment. That they were willing to do so in this cause signaled the importance of their quest. By the second half of the nineteenth century, every self-respecting municipality required deep, reliable wells (or, as in this case, long, reliable pipes). A water supply that was inadequate either in quality or in volume would not only compromise industrial productivity but also reflect badly on the imagination, energy, and competence of the city fathers. For a city such as Manchester, failure to provide such an obviously essential resource would thus undermine both the practical basis of its preeminence and the symbolic status—as icon of industry, trade, and progress—that made it one of the most important places in Britain, and in the world.

Whatever its subterranean anxieties, at surface level the Manchester of the 1870s presented an imposing and assertively self-confident spectacle, the consequence of a century of spectacular expansion. Manchester had been a regional center of cloth production since the sixteenth century. When Daniel Defoe visited early in the eighteenth century he noted that "in consequences of this trade the town has gradually become very large and very populous," comparing it favorably, at least in terms of scale, to such older commercial centers as Chester, York, and Norwich. By 1773 it numbered around forty-three thousand inhabitants. In 1783 an effusive native chronicler, who had transcended his humble beginnings as a fustian shearer to become a schoolmaster, claimed that "the large and populous town of *Manchester* has now excited the attention and curiosity of strangers, on account of its extensive trade, and the rapid increase of its buildings, with the enlargement of its streets." A year later a French observer, whose vision was less likely to be clouded by local pride, characterized Manchester as "superb." The first national census in 1801 counted over seventy-six thousand residents, a number that more than quadrupled by 1851. (The rate of growth of some other industrial cities during this period was similarly rapid, while London's population merely doubled.)[2]

And the population contained within the historical town limits was not the whole story. The official boundary of Manchester moved outward during the Victorian period as the city engulfed and annexed neighboring towns, so that it encompassed five times the acreage in 1913 that it had in 1838.[3] A volume published by the Manchester Corporation in 1894 spelled out the fiscal implications: the "rateable value" of Manchester property approached three million pounds, exceeding "by four-and-a-quarter times the amount at which it stood in 1839." Beyond the city limits population also boomed; the metropolitan area counted over two million residents in 1900, double the number at midcentury. During the parliamentary hearings on the Thirlmere Scheme, the mapped pattern of this urban sprawl was admiringly compared to "the sun in a London fog." Growth was impressive in relative as well as absolute terms. Although Manchester had not ranked among the six largest English cities at the beginning of the eighteenth century, by the end only London outstripped it.[4] According to

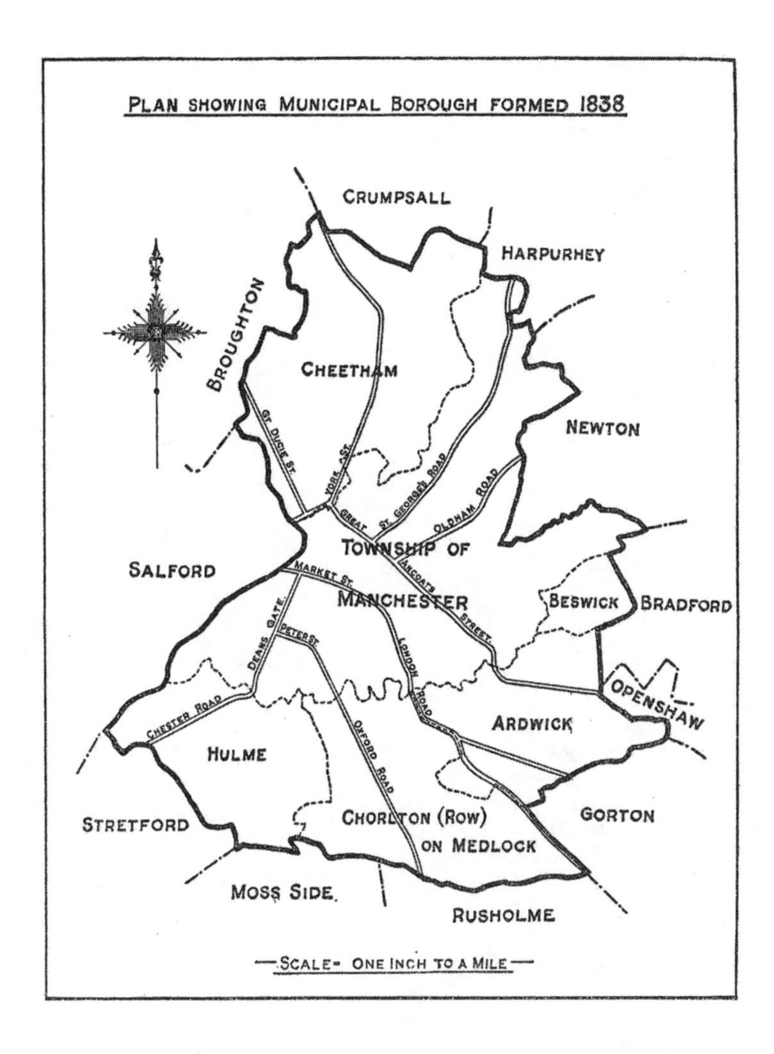

a contemporary guide it had already become, as it was to remain throughout the nineteenth century, "the second town in the kingdom, whether it be considered in a commercial, or . . . in a political point of view."[5]

If captains of Manchester industry and commerce were inclined to celebrate their achievement in reductive statistical terms, they also invested in more substantial and showy testimonials. The cotton mills and warehouses, which lay at the heart of Manchester's prosperity and which constituted its most distinctive and characteristic buildings, took increasingly imposing architectural form. In addition, beginning in the early nineteenth century, prominent Mancunians began to adorn the town center

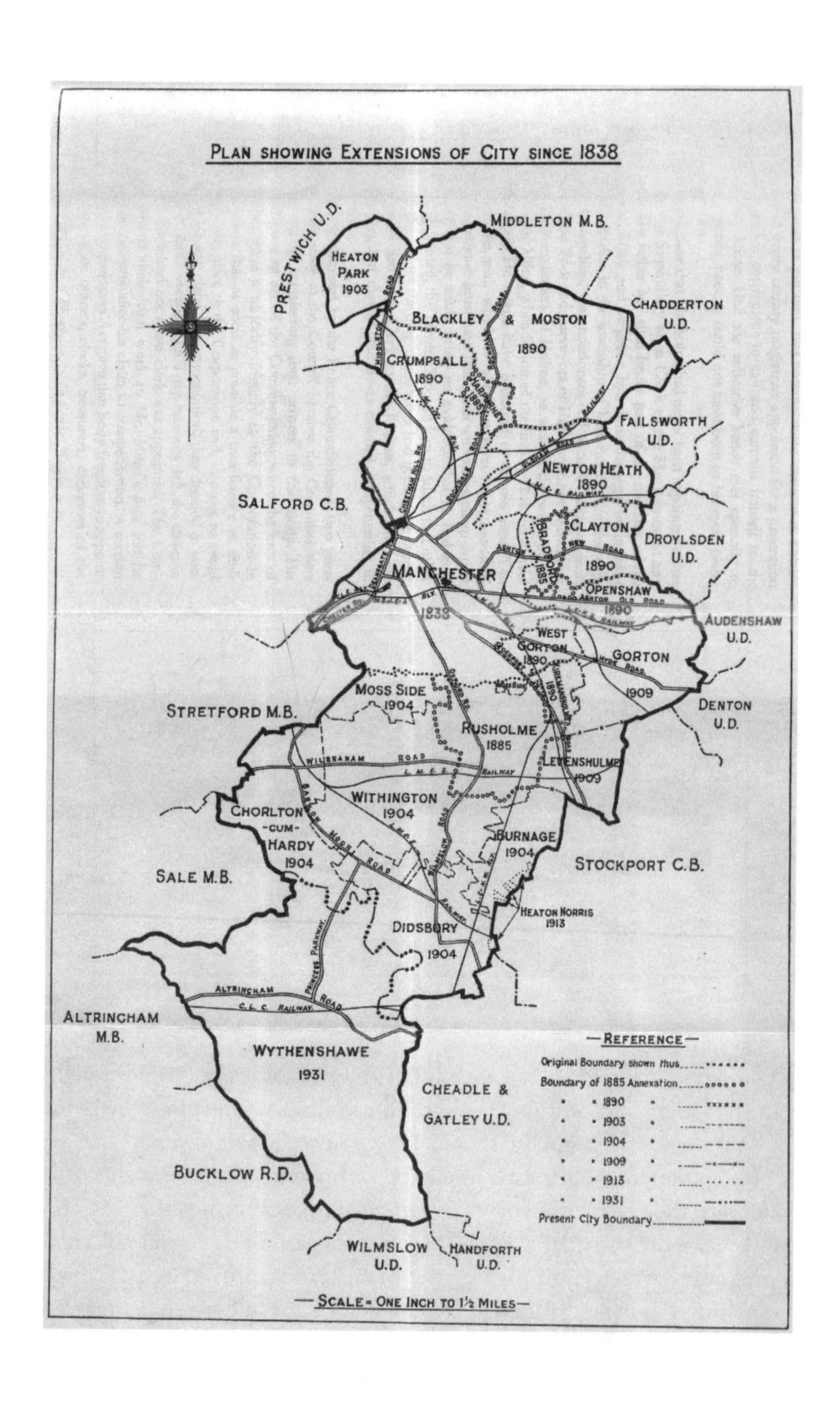

PLAN SHOWING EXTENSIONS OF CITY SINCE 1838
PRESTWICH U.D.
MIDDLETON M.B.
HEATON PARK 1903
BLACKLEY & MOSTON
1890
CHADDERTON U.D.
CRUMPSALL 1890
FAILSWORTH U.D.
NEWTON HEATH 1890
SALFORD C.B.
CLAYTON
DROYLSDEN U.D.
BRADFORD 1885
1890
MANCHESTER
1838
OPENSHAW 1890
AUDENSHAW U.D.
WEST GORTON 1890
GORTON
1909
DENTON U.D.
MOSS SIDE 1904
STRETFORD M.B.
RUSHOLME 1885
LEVENSHULME 1909
WITHINGTON 1904
CHORLTON -CUM- HARDY 1904
BURNAGE 1904
STOCKPORT C.B.
SALE M.B.
HEATON NORRIS 1913
DIDSBURY 1904
ALTRINCHAM M.B.
WYTHENSHAWE 1931
CHEADLE & GATLEY U.D.
BUCKLOW R.D.
WILMSLOW U.D.
HANDFORTH U.D.
—REFERENCE—
Original Boundary shown thus
Boundary of 1885 Annexation
1890
1903
1904
1909
1913
1931
Present City Boundary
—SCALE = ONE INCH TO 1½ MILES—

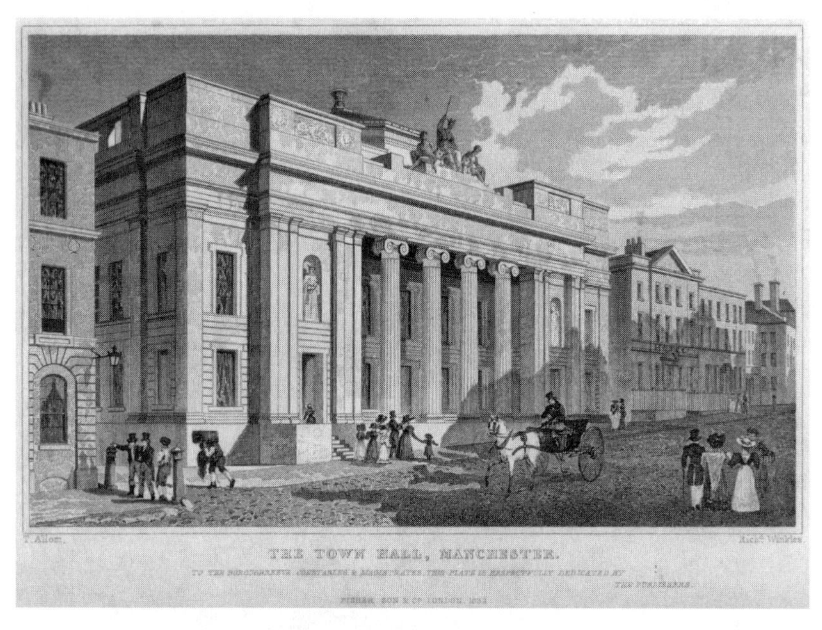

Manchester Old Town Hall, ca. 1832.

with structures emblematic of civic pride and cultural aspirations. Even the classical style of such edifices as the Portico Library (1802–6), the Theatre Royal (1845), the Gentlemen's Concert Hall (1831), the Athenaeum (1836–39), and the Royal Manchester Institution (1824–35, now the City Art Gallery) reflected ambitions that stretched beyond the merely quantifiable. (No mere status symbols, these venues were in frequent use, especially by the city's many cultural and scientific societies.)[6] As Manchester's economic and political power continued to grow, a series of splendid new cultural, governmental, and commercial buildings enhanced its corporate heart. A midcentury guide praised the city's "noble appearance," noting that its "wide, open streets and squares" distinguished it from other "manufacturing towns." By the time of the initial reconnaissance of Thirlmere, the construction of Manchester's most magnificent civic monument, the exuberant Gothic town hall, was nearly finished.[7] The only standard metropolitan distinction that the city still seemed to lack was historical; it had Roman roots, but they did not seem sufficiently robust to support its present vigor and eminence. Perhaps in compensation, the Manchester Royal Jubilee Exhibition of 1887, although primarily devoted to modern technology, also celebrated a romanticized if imprecisely delineated past.

Manchester New Town Hall, 1877.

Visitors could walk through an imaginative reconstruction of the gateway towers of the Roman fort of Mancunium into an anachronistic fantasy of "Old Manchester and Salford" composed mostly of replicas of sixteenth-century buildings. One gratified local journalist referred to the mixture of periods as a "delightful jumble."[8]

The absence of past distinction was, however, only a minor irritation (this was fortunate, as the condition was essentially irremediable). Victorian Manchester was quintessentially a city of the present and the future.

As Asa Briggs famously put it, Manchester was the "shock city" of the mid-nineteenth century, illustrating the extreme exploitation of contemporary commercial and industrial possibilities and forecasting the path that other large manufacturing towns were likely to follow.[9] Shocks, of course, are more frequently unpleasant than enjoyable, so it is not surprising that as it loomed ever larger, both physically and figuratively, Manchester inspired disapproval, dismay, and fear, as well as admiration, especially among those who did not benefit directly from its growth. Some negative responses were predictably reactionary. For example, the influential essayist Thomas Carlyle contrasted a caricature mill owner, "Plugson of Undershot . . . who has indomitably spun Cotton merely to gain thousands of pounds" rather than acknowledging the broader social obligations and ennobling opportunities open to a "Captain of Industry," with the idealized Abbot Samson, the wise, selfless, and autocratic ruler of the twelfth-century abbey at Bury St. Edmunds.[10] Nevertheless, most critics wanted to ameliorate the harshest by-products of progress, not to reject it altogether, and they tended to agree with boosters on several important points. Both groups of commentators were struck by the city's scale, and both understood it as the definitive product of the modern age. From either perspective, Manchester appeared as a representative phenomenon as well as an exceptional one—paradoxically both *primus inter pares* and *sui generis.*

Thus, as advocates praised Manchester in its own right and as a symbol of British progress and prosperity, critics used the same double register to deplore it. In an account published in 1832, for example, James Phillips Kay (later Kay-Shuttleworth) emphasized the awe-inspiring power of "the metropolis of the commercial system" and the awe-inspiring energy of the men who had created it: "A stranger regards with wonder the ingenuity and . . . capacity, which, in the short space of half a century, have here established the staple manufacture of this kingdom. He beholds with astonishment the establishments of its merchants—monuments of fertile genius and successful design . . . and the restless but sagacious spirit which has made every part of the known world the scene of their enterprize."[11] Although his phrases echoed the language of tourist literature and civic boosterism, Kay's breathless description was buried in the middle of a tract with a very different message. As secretary of the local Board of Health, Kay was responsible for organizing Manchester's response to the devastating cholera epidemic of 1832, which took its greatest toll in the sprawling districts that housed the industrial poor. *The Moral and Physical Condition of the Working Classes Employed in the Cotton Manufacture in Manchester* cataloged the crowded and filthy conditions in which most factory workers lived, district by district and sometimes street by street. The preceding pages—filled with graphic descriptions of "houses ill soughed [drained], often ill ventilated, unprovided with privies," streets that "become the common receptacles of mud, refuse, and disgusting ordure," whole neighborhoods whose inhabitants resembled "savages in their appetites and habits," and "factories . . . whose chimneys vomit forth dense clouds of smoke"—provided a sobering and ironic context for Kay's characterization of the achievement of Manchester's "opulent merchants," most of whom chose to reside in more salubrious suburban or rural settings.[12] Writing for a German audience a decade later, Friedrich Engels similarly used Manchester's incontestable greatness to counterpoint its manifest failings and, explicitly as well as by implication, as the basis for his general critique of industrial capitalism. Acknowledging that "the modern art of manufacture has reached its perfection in Manchester," he contrasted the "fine, comfortable homes" and "wholesome country air" enjoyed by the affluent with the mean urban streets, evoked in terms still more lurid than those used by Kay: "everywhere half or wholly ruined buildings . . . heaps of debris, refuse, and offal; standing pools for gutters, and a stench which alone would make it impossible for a human being in any degree civilized to live in such a district."[13]

Water, Water Everywhere

Whether they focused on prosperity or poverty, observers of the Manchester phenomenon agreed on the importance of water. Unlike industrialism, however, water was not a novel component of Manchester history. The modern town had grown up on the banks of three rivers, combining two centers of previous settlement; one near the Roman fort at the confluence of the Irwell and the Medlock, the other surrounding the medieval church of St. Mary near the confluence of the Irwell and the Irk.[14] As was the case throughout the manufacturing belt that stretched across upland Lancashire and Yorkshire, the availability of running water had helped to determine the location of industrial development. By the beginning of the nineteenth century, Manchester's exploitation of its rivers could be quantified as a component of its commercial greatness: the Irk had "perhaps more mill seats upon it than any other stream of its length, in the United Kingdoms," and the Medlock was "highly valuable from its banks being the seat of many dye-houses."[15]

Along with wealth, this profusion brought less pleasant consequences.

As a later chronicler put it (still admiringly), Manchester's "three rivers . . . are all made subservient to the trade of the place, being extensively used for manufacturing purposes." He nevertheless went puckishly on to point out that "the colour of the water in consequence bears strong resemblance to that of the famous river Styx; and the smell at times is anything but agreeable to the olfactory nerves."[16] The manufacture of cotton textiles put especially heavy burdens on the water supply. It required large amounts of pure water, and it returned that water to the rivers in a particularly irretrievable state of chemical pollution, which disinterested observers described with less humor. Johann Georg May, a Prussian civil servant touring England as an industrial spy, noted in 1814 that "the river which flows through Manchester is so filled with waste dye matter that it looks like a dye-vat."[17] Twenty years later, Alexis de Tocqueville similarly suggested that Manchester's "fetid muddy waters, stained with a thousand colours" had become unwholesome open-air extensions of its factories. Extending the metaphor of contamination, he characterized the whole city as a "vile cesspool" out of which, nevertheless, "the greatest stream of human industry flows . . . to fertilise the entire universe."[18]

Such imagery presented the city as a single composite factory. But factories needed workers, and workers needed to live, however marginally. With respect to water, therefore, the interests of Manchester's two great indices of growth—its population and its commerce—ran in different channels. Not only did they make competing claims on supply, but industrial use of river water rendered it unfit for domestic consumption—that is, for drinking, cooking, and washing. (The reverse was also true in theory but not significant in practice.) And so, from the beginning of Manchester's commercial and industrial preeminence, it had a water problem. Indeed, there had been trouble even earlier. For many centuries, demand had been satisfied by abundant soft water from the rivers and abundant hard water from wells drilled into the town's sandstone substrate. By the sixteenth century, however, orders limiting the amount of water that individuals could draw from the principal public fountain suggest that resources were already strained.[19] In the late eighteenth century Sir Oswald Mosley, the lord of the manor of Manchester, underwrote the construction of an engine that pumped water from the Medlock to reservoirs in the town, from which it was distributed through a system of pipes.[20] But burgeoning population and geographical spread, in combination with industrial and organic pollution that drained back into the river and seeped into the water table, rendered this system increasingly inadequate. By the end of the eighteenth century, most households of moderate size, especially in the newer neighborhoods, collected rainwater for domestic purposes in lead cisterns.[21]

Manchester's problems were not unique. Most other urban centers, in Britain and elsewhere in the industrializing world, confronted similar issues of water shortage and water quality. In part this was simply an issue of scale. Thus London, by far the largest city in the nation, had the longest experience in meeting such challenges, and this experience was not particularly encouraging. Sewage disposal had occupied official attention for centuries. A series of acts and regulations, beginning with the Bill of Sewers in 1531, had attempted to cleanse the Thames drainage system, but well into the nineteenth century much of London's water supply remained both offensive and insalubrious. During the same period, a series of private enterprises (both individual and corporate) had conveyed water to consumers, with varying degrees of efficacy.[22] Closer to Manchester, the port city of Liverpool had continued to rely on local springs and shallow wells well into the eighteenth century, although the need to tap larger resources had been recognized by a corporate grant in 1694 and a parliamentary act in 1709.[23] Across the Pennines, the wool-manufacturing town of Bradford suffered chronic water shortages during its early nineteenth-century growth spurt. A small town reservoir enhanced the local supply, but in 1837 most residents still drew their water from wells, springs, or private water carts.[24]

Manchester's early attempts at water solutions fit the general pattern. As it outgrew the small-scale nearby sources of water on which households and businesses had traditionally depended, a patchwork of alternatives emerged. All proved unsatisfactory, for a combination of technological, economic, and political reasons. By the early nineteenth century, the need for a more systematic approach had become obvious. The proprietors of a stone pipe manufacturing concern had purchased the right to furnish water to the town from the lord of the manor in 1807. In 1809 Parliament passed a private bill incorporating them as the Manchester and Salford Waterworks Company (although they were often still referred to as "the Stone-Pipe Company"). There were many local objections to this bill, as well as to a similar bill unsuccessfully put forward by a rival company.[25] At impassioned town meetings, citizens expressed their desire to maintain local control of the water supply, enthusiastically endorsing a resolution that stated, "It would be contrary to sound policy to intrust the furnishing and control of this important article of food and cleanliness, on which the health and comfort of the inhabitants depend, to persons whose sole object will be the promotion of their own private interest."[26] Several generations later, the engineer John Frederic LaTrobe Bateman regretted the failure of the town's efforts at resistance, which had included the expenditure of almost eighteen hundred pounds to lobby against the bill, as well as making speeches and passing resolutions.[27]

Bateman's involvement with the Manchester water supply extended throughout his career. He had begun by consulting for the Manchester and Salford Waterworks Company in its final years, as it struggled vainly to catch up with increasing demand.[28] Nevertheless, like most progressive technocrats of his time, he was an ardent proponent of municipal control. The company had, in his expert opinion, perpetrated "one of the most nefarious and barefaced pieces of jobbery which has ever disgraced the annals of private companies, replete as they unhappily are with instances of dishonesty."[29] Even when described with greater restraint, this enterprise was clearly unsuccessful. The company's initial attempt to use its signature

stone pipes to transport water had to be abandoned when they proved embarrassingly porous, as well as more expensive than iron alternatives. It went into liquidation in 1815 and was unable to establish a sound financial basis for its operations, despite a series of subsequent parliamentary acts. The company's efforts to tap additional sources of water were similarly inadequate. By 1845 it served only about 23 percent of the town, and that portion for only a few hours each day, at rates steep enough to discourage profligate use for such luxuries as washing.[30] On the watch of the Manchester and Salford Waterworks Company, the local rivers had become increasingly filthy; the mood of industrial workers, in Manchester as elsewhere in Britain, had been darkened by economic depression and the Chartist movement; and the inadequacy of Manchester's sanitary arrangements, for both water supply and sewage removal, had attracted official scrutiny and criticism at the national level.[31] Of course, these larger developments lay beyond the power of a private water company to mend or mar, but they intensified the sense of crisis and failure.

Hydration, Representation, and Taxation

The deficiencies of the Manchester and Salford Waterworks Company were duplicated elsewhere, and by the 1840s many towns had become dissatisfied with the performance of private water companies. A water industry run with the short-term objective of maximizing profits could not meet the rapidly expanding needs of urban industries and populations; economic historian J. A. Hassan characterizes the situation as "a classic case of market failure." Municipal ownership allowed a focus on maximizing sales, a longer time frame for planning, and access to both more advantageous financing and larger amounts of cash. For all these reasons, during the second half of the nineteenth century increasing numbers of British towns assumed control of their own water supplies.[32] In most cases the results of these appropriations were highly gratifying. Again, the Mancunian experience was representative. As Shena D. Simon, left-wing politician and former mayor, reflected in 1938, using language borrowed from the liberals of the previous age, "the story of Manchester's water supply is the story of progress." An official publication commemorating the centenary of municipal water control more exuberantly stated that "the provision . . . of an adequate and pure supply of water at an economical price . . . is a notable tribute to the foresight, energy and courage of a succession of public representatives."[33]

If the transition from private to public ownership was impressively steady, it could also be considered surprisingly slow, given the acute need for urban water. Manchester was not the only Victorian town whose core industry was threatened by prospective water shortages, or whose aesthetic and sanitary qualities were impaired by the condition of its taps and rivers. To take the most conspicuous example, London's egregious water problems were addressed only after they had caused serious, if not unpredictable, crises. Contaminated water helped spread the cholera epidemics of 1832 and 1854, and the sewage that choked the Thames caused widespread olfactory distress every summer, culminating in the fabled Great Stink of July 1858.[34] Some general factors that encouraged the persistence of unsatisfactory private or patchwork arrangements were the same for municipalities great and small. Even deficient systems for providing and delivering water were large and complicated, which provided at once a strong theoretical argument for municipalization and a strong pragmatic argument for inaction. Any significant change required parliamentary approval, obtaining which involved both time and expense; compounding the procedural bother was the inclination of early Victorian parliaments to prefer private water enterprises to municipal ones, no matter how vigorously the community to be served expressed a contrary preference.[35] Moreover, geographical units that made sense in terms of water supply tended to have different boundaries than did established units of civic administration, and even within such units, governmental authority was often divided among councils and independent boards.

These common factors were, of course, modified by local circumstances. Manchester's situation struck many contemporaries as particularly difficult and complex—even chaotic. Aggravating the challenges produced by scale were those produced by history. However rich and imposing it had become by 1838, the city, according to Simon, lacked "practically any regulations except those necessary for governing a village or a small town."[36] The reason was that Manchester's urban status, unlike that of the long-established boroughs to which Defoe had compared it, was unacknowledged in law. In consequence it had no formal, unified city government. Sir Oswald Mosley remained the lord of the manor, and administrative machinery designed for a medieval manor had been adapted piecemeal to address the needs of the expanding industrial city. Or at least some of the city's needs; an accretion of supplementary institutions took care of others. A Court Leet, responsible to the lord of the manor, thus coexisted with the Board of Churchwardens and Overseers, which administered poor relief and selected the surveyors of highways, and with the police commissioners, who were responsible for some aspects of town maintenance, including street paving and trash collection, as well as for keeping the peace. Since the members of these bodies were variously appointed or elected, and since they reported to different authorities, they did not feel compelled to work cooperatively.[37]

Manchester did not formally transcend its medieval condition until 1845, when the mayor and corporation paid Mosley two hundred thousand pounds for the manor and manorial rights.[38] But this transaction merely capped a decade of major changes, one consequence of which was the very existence of a mayor and municipal corporation with the authority and resources to make such a purchase. Along with a number of other northern industrial towns, including Sheffield, Bradford, and Leeds, Manchester had lacked independent legislative representation until 1832, when the Reform Act recognized each of them as a borough with the right to elect two members of Parliament. As a result, the voice of the manufacturing interest was greatly strengthened in national debates, and the individual towns had more reliable parliamentary advocates on issues that particularly concerned them, such as the enabling legislation for public works projects. Still more significant with regard to local infrastructure was the Municipal Corporations Act of 1835. Primarily designed to reform the often corrupt or ineffectual corporations that had governed towns like Norwich and York since the medieval period, it also allowed unincorporated towns to petition for incorporation.[39] Desire for municipal status, and the corollary abandonment of existing arrangements, which had shown some recent amelioration, was not universal among the Manchester ascendency. Nevertheless, after a period of predictable struggle between those whose influence was likely to wane in the wake of incorporation and those whose influence was likely to wax, the latter group triumphed, and in 1838 Manchester became one of the first of the manufacturing towns to receive its charter.[40]

Despite the reluctance of the police commissioners, the churchwardens, and the manorial officials to disband, or indeed, to acknowledge the supervening authority of the new body, city council elections were held immediately. At their first meeting the councillors chose an energetic mayor and sixteen aldermen. They also, more importantly as it turned out, appointed Joseph Heron, the son of a local merchant, as town clerk.[41] In the view of some council members this was a bold decision, even a risky one. Despite his record of administrative service during the struggle for the charter and, previously, during the cholera epidemic of 1832, his youth—he was not yet thirty—caused some misgivings. His elegant appearance gave the sober men of business additional pause: "He was a handsome man . . .

THE LATE SIR JOSEPH HERON,
First Town Clerk of Manchester.
(From a Photograph by M. Guttenberg.)

and . . . a gay young spark," a dandy complete with kid gloves and tight boots.[42] He was also a talented lawyer, ultimately becoming one of the preeminent authorities on municipal law, and he soon demonstrated that he was an energetic and imaginative manager. He conceived his official duties in the most expansive possible way, and felt responsible for every aspect of Corporation business, from carrying out the resolutions of the committees through which it did most of its work, to orchestrating its parliamentary dealings. Since his employment was permanent, he soon attained greater practical seniority than most of the elected officials whom he ostensibly served. During his forty-year career, his forthright, efficient management style garnered increasing respect and influence; or, from another perspective, he was repeatedly criticized as bumptious and dictatorial. He came to view himself as the representative, albeit appointed, of the entire city, in contrast to the councillors, who represented particular geographical constituencies.[43]

Although water was a pressing concern of the city as a whole, the city council did not address this issue immediately. There were many other claims upon its early administrative attention, including some last-ditch legal challenges to its legitimacy. So the Manchester and Salford Waterworks Company was left to struggle for nearly a decade after incorporation. It had, after all, made some efforts at improvement. In the mid-1820s, the company had augmented the water supply by building a reservoir at Gorton, about four miles east of the city. Even the critical Bateman grudgingly acknowledged that for a few years afterward, water was "tolerably abundant," at least "as compared with what had previously existed and . . . with other places," but growing demand quickly outstripped the additional supply.[44] By the early 1840s, it had become clear that new waterworks of some kind would have to be built. The company chose to sink a shaft beneath the existing Gorton reservoir, a project which had the advantages of being relatively inexpensive and of not requiring parliamentary approval or the purchase of additional property. It also had significant disadvantages: it would not provide nearly enough additional water, and the water it would provide was hard—suitable for drinking but not for the manufacture of cotton textiles.[45] Any realistic solution to Manchester's water problem would have to be on a much larger scale.

The company had in fact been contemplating such a solution, which was why its directors had sought Bateman's advice. In 1844 he presented them with a report in which he recommended locating an additional reservoir at Longdendale, in the well-watered slopes of the Pennines, beginning about twelve miles east of the city. They found his analyses and projections persuasive and decided to adopt his scheme. Like every large-scale building project, however, it posed political, legal, and economic challenges, as well as technical ones. The targeted area included a number of small woolen and cotton mills, as well as the villages in which their employees lived. They depended on the River Etherow for their water supply—the same water supply that attracted the company. Dealing with the existing occupants and owners proved insuperably difficult, as the company was unable to negotiate a satisfactory agreement to compensate them for their prospective losses. It therefore abandoned the Longdendale plan, and began to look west, toward the Cheshire hills, for an alternative source of water, while several rival companies sought parliamentary approval for their own plans to supply the city.[46] Finally, the Manchester Corporation, liking neither the competing plans nor the loss of control that they implied, decided to take matters into its own hands. During 1846 it arranged to purchase the Manchester and Salford Waterworks Company, and in 1847 the first Manchester Corporation Waterworks Act became law.[47]

After consulting Bateman, the Corporation decided to proceed with an expanded version of his earlier proposal. Longdendale seemed to be made for their purposes. Its waters combined "the sparkling brillancy and taste of ordinary spring water with the softness of rain water," making it "exceedingly pleasant to drink . . . [and] for bleaching and dying . . . unrivalled." The topography was equally attractive. A fellow civil engineer commented that "it would probably be difficult to discover in England many valleys so admirably adapted . . . for the construction of a series of large impounding reservoirs."[48] Bateman was naturally gratified by the Corporation's decision and happy to accept appointment as superintending engineer. But, mindful that the earlier attempt to realize his project had foundered on a monetary rock, he felt that the most important consequence of the Waterworks Act was financial rather than technical. It allowed the Corporation to tax all owners and occupiers of Manchester property in order to defray the expenses of building and maintaining the water supply, a much larger and more reliable income stream than the Waterworks Company had been able to tap.[49] As Bateman anticipated, the Corporation would need to take full advantage of these fiscal powers. Administrative arrangements were consistent with the scale of the undertaking. At first a newly constituted Waterworks and Gaol Construction Committee managed the water supply, but within a few years the municipalized waterworks basked in the attentions of a committee devoted to them alone.

Next to Godliness

When they decided to enter the water business, the members of the Manchester City Council intended, as they indeterminately put it, to build a system of reservoirs "adequate to meet not only the present, but also the prospective wants of the inhabitants of this borough" (as well as those of Salford and several other neighboring towns).[50] Estimating those wants required an elaborate series of calculations based on a problematic series of assumptions. Some of the problems were obviously quantifiable, even if the resulting quantities were based largely on speculation. For example, on the demand side: How fast would the population grow? At what rate would the market for cotton textiles expand? How many new factories would be built? (Manufacturing consumed about one-third of the available water, according to a retrospective modern estimate.[51]) Similarly, on the supply side: How much rain reliably fell in Longdendale? And how large an area would need to be flooded to provide even the ten million daily gallons deemed minimally necessary in 1847?[52] Reducing other problems to numbers involved even more guesswork. The water needs of a family were more difficult to define than were those of a factory. A majority of the potential domestic consumers of Manchester water lived in the notoriously unsanitary poorer districts of the city. Since even the most parsimonious Manchester politicians aimed to provide more than the minimum amount of water necessary to sustain human life, estimating probable or desirable consumption entailed subjective judgments about appropriate levels of hygiene, comfort, and convenience. (The people setting these targets, it might be noted, were not likely to have their own usage constrained by such estimates.)

The difficulty of calculating domestic water usage did not, of course, mean that estimates were unavailable. Because they were a necessary component of the planning and authorization process, they had to exist. The surveyors who reported to the Commissioners of Her Majesty's Woods, Forests, Land Revenues, Works, and Buildings on a supplementary Manchester Waterworks Bill in 1848, assumed a daily requirement of twenty gallons per head.[53] This was a standard allowance; it reappeared, for example, in the authoritative prescriptions that John C. Thresh offered at the end of the nineteenth century. Both a sanitary engineer and a doctor, Thresh wrote *Water and Water Supplies* in order "to place within the reach of all persons interested in public health the information requisite for forming an opinion as to whether any supply or proposed supply is sufficiently wholesome and abundant." He calculated that "a cleanly man belonging to a fairly cleanly household" would require between fifteen and twenty gallons of water each day.[54] This precise formulation was supported by an analysis of daily usage into its components (drinking, washing, and so forth). But if the definition of a gallon was beyond question, the terms "cleanly" and "fairly cleanly" were open to interpretation. And even the numbers could be challenged. In 1886 an editorial in the *Health Journal* queried the assumption that twenty gallons per person per day would suffice "for all the purposes of domestic life," citing "the highest authorities" in support of the claim that at least 50 percent more would be required. It specifically compared Manchester's stinginess in this regard with the relative liberality of London, Glasgow, New York City, and ancient Rome.

(Ten years later Thresh estimated that Manchester provided only fifteen gallons.)[55]

The largest variable in Thresh's breakdown related to water closets, which added, in his estimation, between two and six gallons to each individual's daily usage, depending on the flush mechanism—that is, up to one-third of the total.[56] Water closets were also the only item on his list that he flagged as discretionary. Although, like many other conveniences, water closets had ancient antecedents, they did not assume their modern form until the nineteenth century. During that period they became increasingly common as domestic fixtures, and their design was continually improved. From the beginning, they loomed large in Manchester water policy debates. When critics like Kay or Engels described the shocking conditions in which the industrial poor lived, the profusion and ubiquity of human excrement inevitably provided the most arresting details. This situation reflected both the inadequacies of the privies (or earth closets, as they were alternatively termed), in number and in condition, and the insufficient municipal arrangements for removal of waste. In 1844 the city council appointed a Building and Sanitary Regulations Committee to deal with this problem (among others). It chose the path of least resistance, requiring the provision of additional privies and prescribing higher standards in their construction. This decision produced some amelioration, at least during the committee's initial flush of energy. In its first year of operation, it examined the dwellings occupied by about one-fifth of the city's population. As a result, the residents of those neighborhoods enjoyed one privy for every three dwellings, or twelve people; previously, the average privy had served twelve dwellings, or sixty people.[57] For various political and legal reasons, the committee could not maintain this rate of transformation. And in any case Manchester's massive sanitation problems could not be solved one privy at a time.

The path of least resistance was also the path of least expense. Unlike every other category of domestic water usage, water closets required that water be piped into (or at least very close to) individual dwellings. Neither the city council nor its middle-class ratepayers looked favorably upon the additional expense that the provision of such elaborate additional infrastructure would require. Thus, although water mains had been installed in all parts of the city by 1852, and piped water was available in the middle-class suburbs, the provision of internal water in working-class neighborhoods did not become official policy for four more decades. This long-delayed change constituted a reversal of the previous policy of explicit discouragement and refusal, which had reflected not only parsimony, but deep indifference (or worse) to the living conditions of the town's underclass.[58]

In addition, complacent estimates of the number of gallons available per person per day normally neglected to specify how much effort individuals would have to expend in order to gain access to their share. Water for household use had to be fetched from common pipes, and weary haulers could be tempted to avail themselves of streams, and even puddles, that were more convenient if less wholesome.[59] Maintaining personal hygiene also required energy and enterprise. If affluent Mancunians had declined to pipe water into the homes of "the poorer classes," they could congratulate themselves for having, in 1845, charitably raised the money, "partly by subscriptions and partly by the proceeds of a grand fancy ball," to establish public baths and washhouses. In these institutions a tepid bath was available for twopence, and a "better class bath" at treble that charge; there was also a small fee for washing clothes. For these modest prices, according to an admiring report, each bather and launderer was provided with a fresh tub of water. (Nevertheless, there was official concern that youthful bathers might save money by using the polluted local canals.)[60] Much later, in 1877, the city council decided to take responsibility for popular cleanliness, at least to the extent of providing public bathhouses in the most densely settled neighborhoods, some of which offered facilities for swimming as well as for washing. After a few years a guidebook to the city noted the enthusiasm with which they were patronized, singling out the swimming contests for special praise.[61] The cost of constructing these bathhouses (less than fifty thousand pounds for the most expensive) was relatively small, a drop in the bucket of Manchester's expenditure on waterworks. The council reconciled itself to the outlay with "the conviction that any pecuniary loss . . . would be more than compensated for by the increased comforts of the people, and by the assistance which would be rendered to the various departments of the Corporation in their efforts to improve the sanitary condition of the City."[62]

The city council had long resisted the insight that the greatest benefit, with regard to the sanitary condition of the city, would derive from making indoor plumbing and water closets available to all residents, irrespective of class. Councillors were, however, acutely aware that besides extra expense, such an expansion would have produced additional river pollution and additional burdens on the sewage disposal and treatment system. Nonetheless, the issue surfaced again and again, indicating that support for such action was, if not constant, at least recurrent. Indeed, it was an ineluctable feature of every discussion about expanding Manchester's waterworks, but until the very end of the nineteenth century, arguments about cost and class repeatedly led to the same minimalist conclusion. During the hearings of the parliamentary Select Committee on the Manchester Corporation Water Bill, in 1878, this official restraint was cited as evidence of the prudence and modesty of Manchester's estimate of its own water needs. The town clerk (by then he had become Sir Joseph) testified that, to this end, it was necessary to suppress the installation of baths and water closets in what he termed "small houses."[63] Sir Edmund Beckett, the distinguished barrister who introduced Manchester's case before the select committee, raised this policy to the level of an essential "peculiarity" or distinction of the place, even a matter of principle or at least of fiat. Manchester was not, he explained, "what is called a water-closet town"; it was rather an earth-closet town. As was the case with many principles and fiats, struggle was required for its maintenance. The policy involved, he regretfully reported, continual struggle against "the popular will. People like water-closets."[64]

Longdendale

Work on the Longdendale reservoirs began immediately after the city received parliamentary authorization. Bateman oversaw all aspects of planning and construction, although much of the practical engineering fell to his former pupil George Henry Hill. Their labors (and those of many construction workers) quickly bore fruit. Longdendale water flowed into Manchester in 1851, only four years after the project was authorized, and the initial daily volume of over ten million gallons increased steadily over the succeeding decades as the entire valley was gradually filled with a series of connected reservoirs.

Although in comparison to subsequent water projects in Britain and elsewhere, the scale of the Longdendale reservoirs seems modest, when they were designed there existed no models of equivalent size; as early as 1852 the *Manchester Guardian* claimed that they had "the largest aggregate capacity of any artificial sheets of water in the world." The Longdendale waterworks remained one of the major achievements of Bateman's distinguished career, which included numerous projects throughout the United Kingdom, as well as consultancies as far afield as Istanbul and Buenos Aires.[65] In addition, since he did not pronounce them completed until 1877 (additional construction continued until 1884), the Longdendale waterworks constituted a kind of leitmotif connecting the successive periods of his life. His monumental *History and Description of the Manchester Water-*

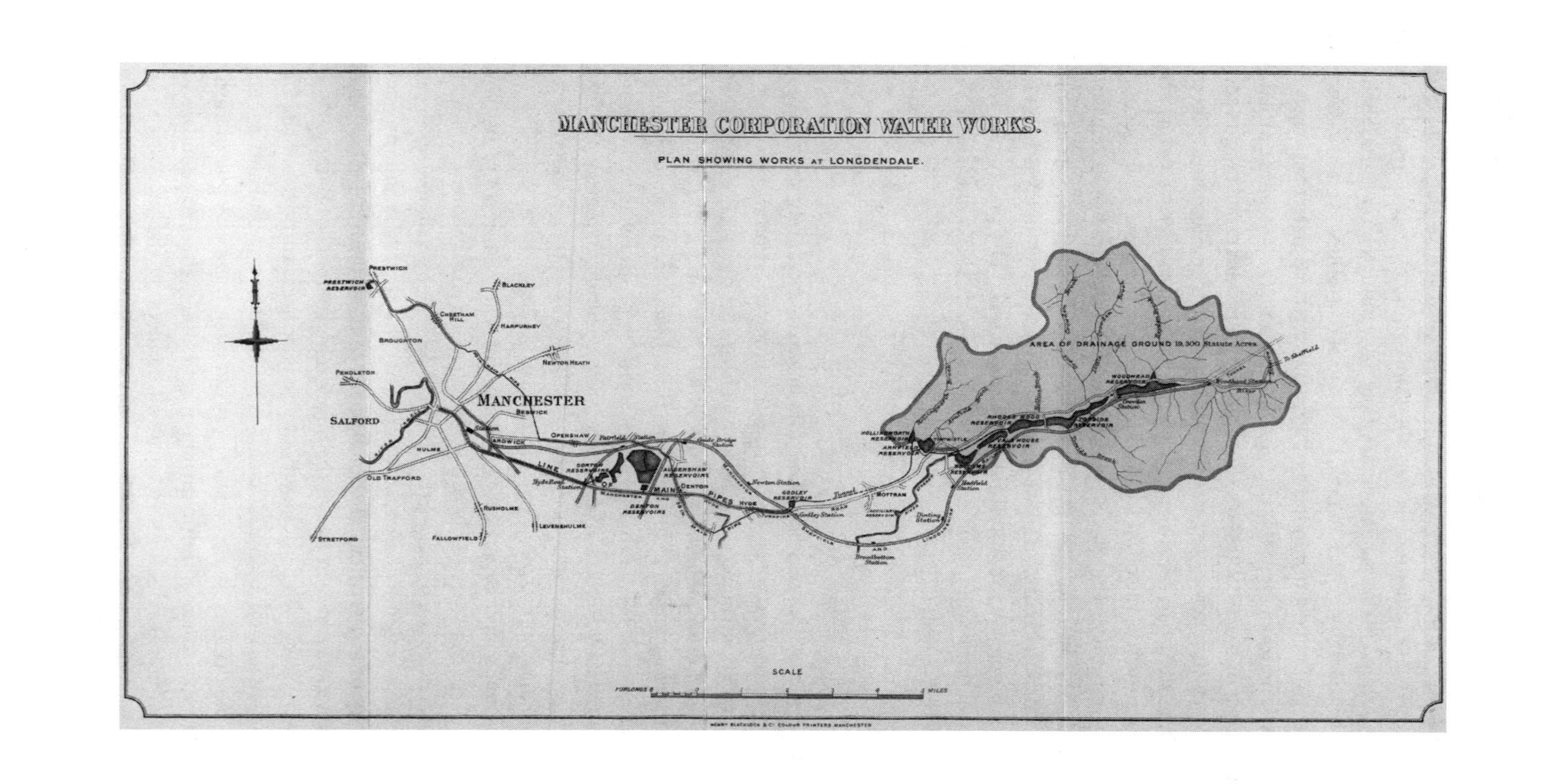
MANCHESTER CORPORATION WATER WORKS.
PLAN SHOWING WORKS AT LONGDENDALE.
PRESTWICH
PRESTWICH RESERVOIR
BLACKLEY
CHEETHAM HILL
HARPURHEY
BROUGHTON
NEWTON HEATH
PENDLETON
MANCHESTER
SALFORD
BESWICK
ARDWICK
OPENSHAW
HULME
OLD TRAFFORD
RUSHOLME
LEVENSHULME
FALLOWFIELD
STRETFORD
GORTON RESERVOIRS
AUDENSHAW RESERVOIRS
DENTON
DENTON RESERVOIRS
LINE OF MAIN PIPES
HYDE
Newton Station
GODLEY RESERVOIR
Godley Station
Tunnel
MOTTRAM
Broadbottom Station
Dinting Station
Hadfield Station
HOLLINGWORTH RESERVOIR
ARNFIELD RESERVOIR
TINTWISTLE
VALEHOUSE RESERVOIR
RHODES WOOD RESERVOIR
TORSIDE RESERVOIR
WOODHEAD RESERVOIR
Crowden Station
AREA OF DRAINAGE GROUND 19,300 Statute Acres
To Sheffield
SCALE
FURLONGS 8
MILES
HENRY BLACKLOCK & Cº. COLOUR PRINTERS MANCHESTER

works, adorned with portraits of Bateman at various ages, as well as maps and plans of the reservoirs at various stages of construction, therefore commemorated the engineer as well as the project. Bateman's retrospective summary of his own contribution might seem somewhat generous: "I have been mainly instrumental in conferring on the inhabitants of Manchester and the neighbourhood one of the greatest blessings which it is possible for a town to enjoy."[66] But however responsibility for this achievement might be apportioned, the value of the Longdendale reservoirs was widely appreciated. The *Manchester Guardian* announced in 1877 that "the public will learn with satisfaction that the last finishing touch has lately been given to the gigantic works in the Valley of Longdendale." A year later, the Waterworks Committee accorded them the highest possible praise, noting that "notwithstanding the additional and enlarged works and consequent increased expenditure . . . [they were] the cheapest gravitation waterworks in the world."[67]

As Bateman himself wrote, the fact that the extensive and elaborate waterworks were "in many respects the largest that have ever been executed in this country" had been cause for anxiety as well as celebration. It meant that the project had been attended with "many difficulties in . . . construction" at the same time that it presented "many points of novelty and interest."[68] The siting, design, and construction of the reservoirs and of the pipes that would convey the water to Manchester stretched existing science and technology to their limits, and sometimes beyond. For example, fundamental considerations in choosing a location for a reservoir system was the quality and quantity of available water. With regard to quality, chemical analysis persuasively supported the encouraging evidence of the senses. Longdendale water was considered pure enough that the system was originally constructed without treatment facilities, and this early analysis stood up robustly to subsequent examination. In 1874 even a chemist who considered open reservoirs to be insanitary could find "only a small amount of . . . organic matters, animal and vegetable, chiefly larval," when he analyzed a sample of water from the tap in the first-class refreshment room of one of the main Manchester stations (he had hoped and expected to find more and worse).[69] With regard to quantity, however, predictions were less precise. It was clear that the Longdendale valley received a great deal of rain, even more than nearby Manchester. According to Bateman, it was "so situated as to be the first to interrupt the progress of the clouds borne by the westerly winds from the most rainy quarters." Exactly how much, however, turned out to be impossible to specify, because there existed no sufficiently detailed and reliable records of local rainfall.[70]

At least the engineers were aware of the deficiencies in their meteorological data. Information that was precise but erroneous could be much more problematic. The authoritative description of the geology of Longdendale on which the original designs had been based proved to require significant revision. A series of accidents revealed that the underlying structure of the valley was much less stable than had been initially assumed. The beds of soft shale that were intermingled with harder rock had caused numerous landslides, the evidence of which was often concealed beneath the ground, and would continue to cause more. As a result, the foundations on which the reservoirs rested had an unnerving tendency to shift, producing technical problems that the initial plans had not been designed to solve. Cracks in the embankments had to be plugged frequently, as part of a continual struggle to render the reservoirs watertight.[71]

And leaks were far from the worst of the engineers' worries. The increasing size of reservoirs sparked public anxiety about their safety, a concern that was heightened in the Pennine region by the disastrous failure of the Bilbury Reservoir near Holmfirth, Yorkshire, just on the other side of the hills. Nearly a hundred people died as the reservoir waters raced through the town in February 1852. The unusually heavy rainfall that triggered the collapse of the Bilbury Dam also strained the capacities of the Longdendale reservoirs. As the *Manchester Guardian* reported, "The anxiety was of course greater than would otherwise have been the case, from the vivid remembrance of the recent horrors attending the bursting of the Bilbury Reservoir . . . though there is no analogy in the nature and character of the works to warrant any apprehension of the one because the other had failed, as for years past all conversant with it had expected it to do."[72] Such reassurance was only moderately persuasive, as evidenced by the persistent fears of local residents and by the arrival on the scene of a throng of ghoulish sightseers, hoping to witness a second catastrophe. There was, after all, substantial basis for their hopes and fears. Although the reservoirs held on this occasion, Longdendale had already witnessed a significant failure in 1849, when a weir at one of the reservoirs gave way, and it was to do so again in 1854, when the embankment of another slid sufficiently to break its discharge pipes, requiring a two-year closure for repairs.[73] The constant possibility of such incidents made the engineers' task much more difficult and stressful. By 1858 Bateman hoped, at least, that similar geological miscalculations might be avoided in the future: "Perhaps the experience I have now gained, and the eye I now claim to have for a land-slip, might enable me to detect difficulties which I could not then [when the plans were originally drawn] perceive."[74]

Naturally these technical challenges also concerned the Waterworks Committee and the city council as a whole. The solution of each unanticipated problem required unbudgeted expenditures, distressing in themselves and the more distressing because delays in the construction of the reservoirs inevitably led to delays in the delivery of their contents to Manchester. And when the water finally arrived, there was often less of it than had been promised, since worries about Longdendale's geological instability made the engineers unwilling to fill the reservoirs to their projected capacity.[75] Time, money, and managerial attention were also consumed by less fundamental issues. There were occasional disputes with laborers over wages; for example, in 1868 a special subcommittee of the Waterworks Committee recommended that stonemasons working at Longdendale "be paid as high a rate of wages as may be paid by other employers in the neighbourhood." Similar trouble could also emanate from the other end of the chain of command. Bateman made intermittent requests for additional compensation, and on one occasion he demanded arbitration and threatened to quit.[76] A more substantial and continuous drain was the obligation to provide compensation water to local mill owners, to replace what they had previously abstracted from the River Etherow, a burden that had previously proved too heavy for the defunct Manchester and Salford Waterworks Company. For several decades after the opening of the first Longdendale reservoir, more water was delivered to these mill owners than to consumers in the city. Since compensation was calculated on the basis of estimated rainfall, this excessive obligation constituted one very practical consequence of the initial inadequacy of meteorological data.[77] The final cost of the project was over £2,500,000 (over £3,000,000 if the cost of purchasing the original company works was included), many times more than the approximately £100,000 that Bateman had estimated when he proposed what he retrospectively termed "the germ of what was subsequently carried out" to the directors of the Manchester and Salford Waterworks Company in 1844.[78]

When the Longdendale project was completed, it provided approximately twenty-four million gallons of water per day, an enormous increase over what had previously been available.[79] It had supported the continuing growth of Manchester's population and economy for more than three decades, and it had produced many specific benefits. For example, the additional water, liberally distributed to hydrants in warehouse districts, and the high pressure at which it was delivered, greatly enhanced fire control. Fires became much less destructive, as measured by the value of property lost, and it became financially prudent to build taller warehouses. Bateman proudly reflected that "streets which were formerly cottages, or small houses of two storeys . . . , have been converted into streets of warehouses, four or five times as high as the buildings destroyed."[80] Per capita water consumption rose, and with it the general level of domestic amenity, as industrial workers availed themselves of public baths and suburbanites watered their gardens and their horses.[81] Although the Longdendale reservoir project did not offer the Corporation much opportunity for permanent commemoration of its achievement, at least not at the site of consumption, visitors to the valley (an excursion recommended by a contemporary guide to Manchester) could admire the valve house at the Bottoms Reservoir. A nearly cubical gray stone structure in severe classical style, it was described by Bateman as an "ornimental Temple," to be adorned on the outside with a granite plaque listing the parliamentary acts that had enabled the Longdendale works along with the chairmen and deputy chairman of the supervising committees, and on the inside with a marble plaque listing every member of the Waterworks Committee from 1846 to 1876.[82] And once there, tourists could admire the impressive scale of the reservoirs themselves, stretching for eight miles along the bottom of the scenic valley.

Yet great as it was, the accomplishment represented by the Longdendale project proved insufficient. Perhaps it was not surprising that the reservoirs failed to address many of the problems so vividly described by Kay and Engels. Because of the differential distribution of water according to class, and the general reluctance of the Corporation to invest in sanitary reform, neither public health nor the urban environment improved significantly.[83] The Corporation had wholeheartedly invested in waterworks, in the expectation that its massive expenditure on Longdendale would secure an adequate future supply. When the reservoirs were completed and as full as possible, however, they held barely enough water for the city's current needs. The city had grown more rapidly than had been anticipated in the 1840s. And individuals were using more water. The availability of new supplies had fueled new demand; as the chairman of a later Waterworks Committee put it, "Such . . . was the effect created by the supply of the pure and wholesome water . . . , as compared with the unsatisfactory supply previously given, that demand at once began to largely extend, being by the end of 1864 about three times as much as it was in 1851."[84] As Bateman celebrated the completion of his Longdendale labors in 1877, therefore, future shortages loomed. Although the *Manchester Guardian* "presumed that the water supply . . . will be satisfactory, at least for some few years to come," other voices were less confident. Only a year later the civil engineer

James Mansergh warned that within a decade the existing supply would barely suffice to meet the needs of the city during a drought, when the water level in the reservoirs would be low.[85] His prediction proved extremely accurate.

Bateman, who possessed the most thorough understanding of the promise and the limitations of the Longdendale reservoirs, had long feared that they would prove the beginning, rather than the end of the municipal search for water. Starting in 1855, he had repeatedly warned that Manchester's growing needs were likely to outrun the Longdendale supply, and in the drought year of 1868 he reported to the Waterworks Committee that, assuming a reasonable increase in consumption, "you have not more than about seven years supply . . . , by which time you must have obtained additional supplies of water if the . . . prosperity of the district [is to] be secured by the abundance of water for personal comfort and for the wants of trade." By 1875 his anxiety had become acute and his rhetoric less formal. He began a letter to the town clerk, addressed simply to "My dear Heron,"

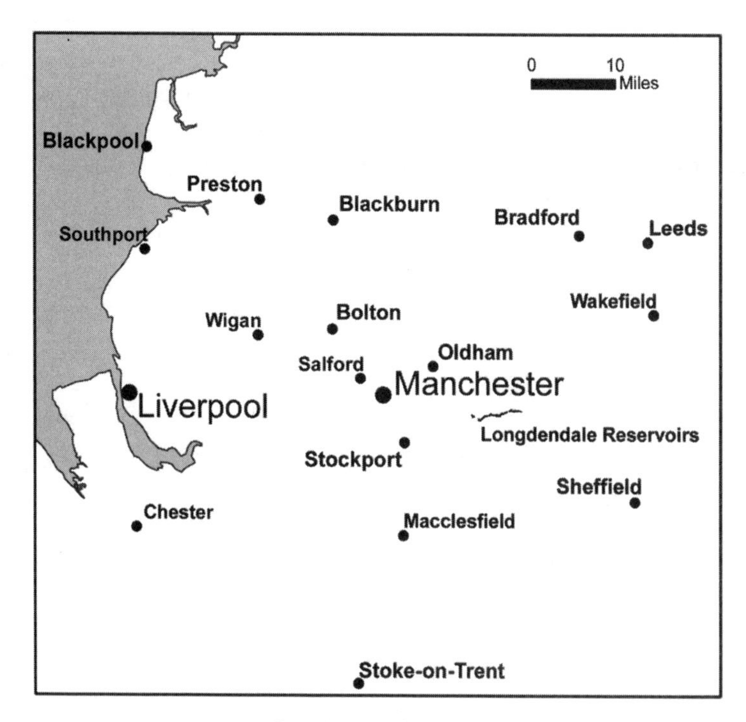

Competition for water.

with an abrupt statement of haste: "I wish I could impress you with the importance of losing no time in the endeavor to obtain additional supplies of water." He forecast that the water might run out in as little as four years. He characterized this eventuality as a "terrible calamity" and an "impending catastrophe" and professed himself ready to "do anything to avert it." Indulging in a bit of self-pity, he regretted he had been "foretelling this state of things for years, but, I am afraid, to many unwilling ears."[86]

The tide, however, was already turning. The need for a substantial supplement to Manchester's water supply was soon widely, if not universally accepted. When Mansergh offered his prognostication several years later, he was part of a growing chorus. Like some other members of this chorus, including Bateman, the independence of his judgment was open to question, since the Manchester Corporation had retained him to testify on the city's behalf before the parliamentary select committee. But the need for more water, although still the subject of political argument, did not, by this time, provoke much technical controversy. Open to greater debate was where to find it. There were no suitable sources remaining nearby. Even in the 1840s, the construction of the Longdendale reservoirs had required expensive and time-consuming negotiations with existing owners and occupants. In the intervening decades, the entire region had become more densely developed. Manchester was surrounded by other towns and cities, whose expanding populations and factories were equally eager for water, and all likely reservoir sites in the Pennines—at least all that were of sufficient size—had already been appropriated "by some large town or company."[87] So the Manchester City Council had to look further afield—as it turned out, much further afield, to the rainy Lake District one hundred miles to the north.

3

The Struggle for Possession

It was an open secret that the Lake District was full of water. Although the Cumbrian lakes had quenched only local thirsts before Manchester looked in their direction, they had already been recognized as a valuable national resource. In 1866 the Royal Commission on Water Supply had been charged with "ascertaining what supply of unpolluted and wholesome water can be obtained by collecting and storing water in the high grounds of England and Wales, either by the aid of natural lakes or by artificial reservoirs." The commission's first priority was "the supply of the Metropolis and its suburbs," but it was also instructed to consider the needs of the other "principal towns."[1] So seriously did the members of the commission take the possibility of a remote supply for London that the first section of their report dealt with proposals to pipe water from the sources of the Severn and the Wye (both in Wales), the hills of Derbyshire, and, most distant of all, the hills of Cumberland and Westmorland. The Lake District plan, as projected by engineers George Hemans and Richard Hassard, was elaborate as well as extensive, involving three of the larger

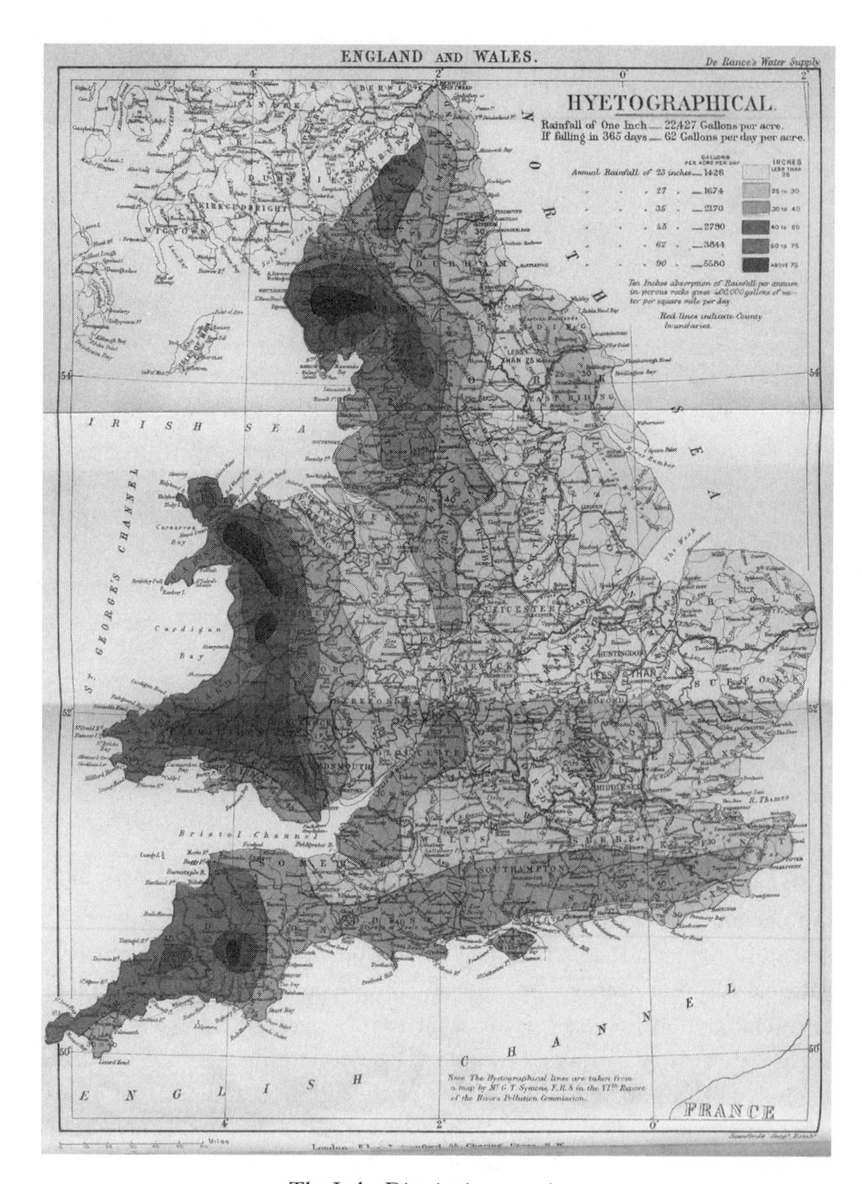

The Lake District is very rainy.

Lake District rock is not porous, so most rainfall runs into the lakes rather than being absorbed.

lakes. Thirlmere and Haweswater were to be dammed and connected by pipeline to Ullswater, from the southern end of which a conduit would run approximately three hundred miles to London.[2] The commission considered the plan "practicable" in principle, despite posing significant technical and financial challenges. In the end there was no need to rise to these challenges, as an adequate water supply for London turned out to be available much more conveniently and economically within the Thames drainage system. But the commission offered a different reason for not pursuing the northern option: "This district is not unlikely to be claimed as the most natural source of supply for large and increasing manufacturing populations in the north of England . . . and we hold it to be erroneous in principle that any one town or district should take possession of a gathering ground geographically belonging to another, unless it can be clearly shown that circumstances render such a step justifiable."[3]

The members of the Waterworks Committee of the Manchester City Council shared this sense that moral claims could be asserted on remote territory that they did not own. They had proceeded with ponderous deliberation toward the decision that a new water supply was necessary. For years Bateman's dire prognostications had fallen on deaf (or deafish) ears. In October 1874 he once again warned a subcommittee constituted to consider "Supply of Water and Proposed New Works" that "the daily increase in . . . consumption . . . necessitated further provision . . . in addition to the water obtained from the present works in Longdendale."[4] On this occasion, his audience proved responsive, and a mere five months later the entire Waterworks Committee concurred "in the desirability of obtaining a further supply of water."[5] Once persuaded of the need for significant additional resources, the thoughts of the committee turned ineluctably to the north, where, according to Bateman, "the vast quantity of water precipitated on the Cumberland and Westmorland hills . . . was running uselessly to waste."[6]

Their only question was which of the many Cumbrian lakes would best serve Manchester's purposes. Their answer was determined on the most pragmatic possible grounds. The Waterworks Committee considered each of the lakes earlier proposed for the London supply, since these alone combined large size with the high elevation that would allow gravity to do most of the work of transportation. At first Bateman recommended Ullswater, the largest of the three, in the hope that Liverpool, which was also short of water and which had also requested his advice, would collaborate with Manchester on a shared reservoir and pipeline.[7] Fearing the "complicated arrangements" that such cooperation might entail, however,

Liverpool decided to pursue an independent water supply. Ultimately it constructed its own reservoir by damming the River Vyrnwy in Wales.[8] Although Bateman continued for a time to urge Ullswater as the best location for a solo Manchester scheme, the city council decided instead to follow the advice of Alderman John Grave, who was chairman of the Waterworks Committee as well as a former mayor and a native of the Lake District, and set its sights on Thirlmere.[9]

Thirlmere was not as large as Ullswater, but, as Bateman reported in 1877, the fifty million gallons a day it was expected to supply would probably be sufficient "for the next 30 or 40 years or longer" (as it turned out, an accurate prediction).[10] Its elevation, sixty feet higher than that of Ullswater, was to its advantage. (Although Haweswater lay higher still—694 feet above sea level to Thirlmere's 533—it received 30 percent less rainfall.)[11] And Ullswater had other drawbacks, including the many private residences scattered along its shoreline (which were, as a chronicler of the completed Thirlmere Scheme later put it, "valuable in the opinion of the owners") and pollution from the substantial Greenside lead mines near its southern end, where the pipeline would originate.[12] The two lakes were about the same distance from Manchester, but a pipeline from Ullswater would require an eight-mile-long tunnel, which would be much more

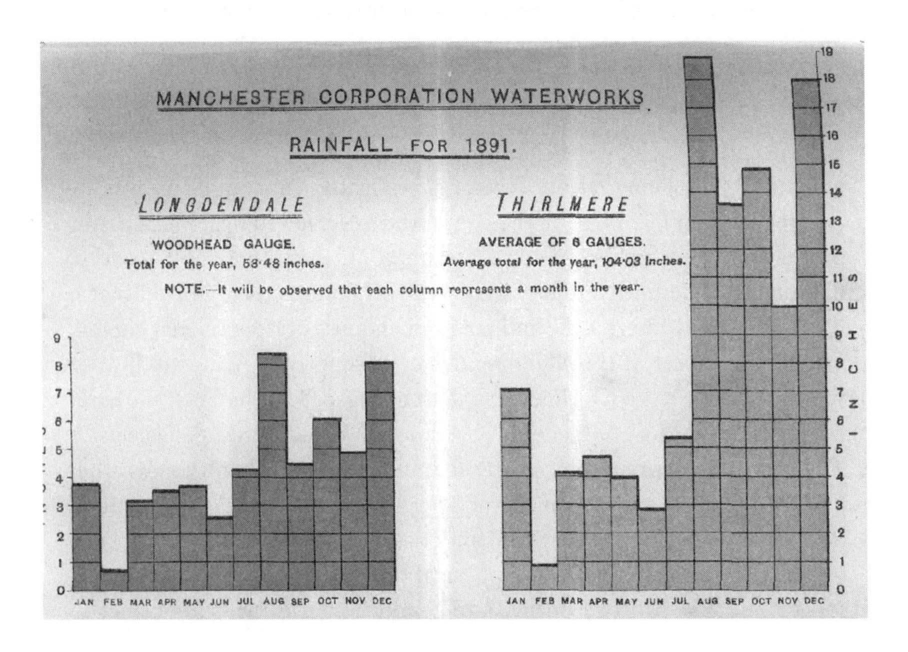

expensive to build than the short tunnel through Dunmail Raise, the hill at the southern end of the valley of Thirlmere, that would be needed to transport that lake's water. Thirlmere's harsh geological context, which produced its sublime and dramatic appearance, also enhanced the quality of the water. The relative infertility of the soil meant that little vegetable matter, whether from plowed fields or from wasteland, washed into the lake. The steep surrounding cliffs did not absorb rain, so the lake was likewise uncontaminated by leached mineral deposits: "the consequence is that even after the heaviest storms the lake is never turbid, and the water is bright, pure and delicious." (Of course the Waterworks Committee members did not judge the quality of Thirlmere's water solely on the evidence of their senses; the analyses of several eminent chemists confirmed its purity and freedom from "animalcule and vegetable life.") Finally, at least by Manchester standards, the area surrounding Thirlmere was "almost entirely destitute of population."[13] In all, as the bishop of Manchester was reported to have exclaimed on viewing the site of the proposed reservoir: "If Thirlmere had been made by the Almighty expressly to supply the densely populated district of Manchester with pure water it could not have been more exquisitely designed for the purpose."[14]

The Properties of Water

The members of the Waterworks Committee were unwilling to make a decision with such enormous potential consequences for Manchester's sanitary and financial well-being on the basis of hearsay, even from such reliable sources as Bateman and Grave. But most were active men of business, and once they had made a decision, they lost no time in laying the groundwork for its implementation. Thus an investigative subcommittee was appointed, and on July 13, 1876, the committee resolved to begin purchasing property in the neighborhood of Thirlmere "if it is practicable to do so upon reasonable terms." On August 2 the subcommittee reported that it had "visited Thirlmere Lake and made all enquiries respecting its eligibility and the extent of the adjoining owners' property."[15] Although in later years official visits to Thirlmere came to be regarded as junkets, the earliest investigations were "generally arduous, and sometimes dangerous."[16] The rugged Cumbrian topography and unpredictable Cumbrian weather provided much of the discomfort and peril, since their task often required that the Manchester emissaries abandon the few good roads and proceed by rugged hill footpaths. These physical difficulties were exacerbated by social ones. The prospectors' wish to conduct their initial surveillance in secret limited their ability to engage local assistance. Their primary motive for stealth was financial. They knew that as soon as their intentions became public local land prices would jump, and they wished to purchase as much property as possible before the inevitable inflation. Unsurprisingly, however, their cover was soon blown; detailed curiosity about local land holdings was unusual, and most committee members were apparently easy to distinguish from the ordinary residents of rural Cumberland. In response to a suggestion, during a much later arbitration, that "a good many of the persons who sold to you did not know that you were acting for the Manchester Corporation," the aggressively energetic John James Harwood, who had been a member of the first exploratory subcommittee, testified, "Oh, I think they did, because it was commonly known in the neighbourhood."[17]

The Waterworks Committee was interested in controlling the entire drainage of Thirlmere, an area that totaled approximately eleven thousand acres. The continuing need to negotiate with abutting landowners at Longdendale, and the constant worry that anyone with access to the streams that fed the reservoirs might contaminate the water, had convinced committee members of the extreme desirability of eliminating competing claims and influences in their Lake District project.[18] Their most important initial targets were the two largest estates, which together occupied much of the lakeside and the surrounding uplands. Much of the watershed fell within the manor of Wythburn, owned by the Vane family, whose seat was at Hutton Hall near Penrith, about twenty-five miles to the east. The head of the family was Sir Henry Vane, although at the time of Manchester's initial approach his right to that position was under legal challenge. In a Chancery suit that had begun in 1872, his uncle Frederick Vane claimed that his older brother (Sir Henry's late father) had been born immediately before their parents' marriage rather than immediately after it, and that Frederick was therefore the first legitimate son. The case was decided, resoundingly in the incumbent's favor, in late November 1876, by a judge who clearly found the unfilial airing of even potentially dirty linen distasteful.[19]

Apparently Sir Henry had entertained no serious doubts about his father's legitimacy or his grandmother's virtue, at least none serious enough to prevent him from opening negotiations that would lead to the disposal of a large piece of his patrimony. By early September the subcommittee minutes recorded (somewhat prematurely) that he had agreed to sell the manor of Wythburn along with all associated rights to the Manchester Corporation for twenty-five thousand pounds, about 20 percent more than the value that had been estimated on the basis of previous surveil-

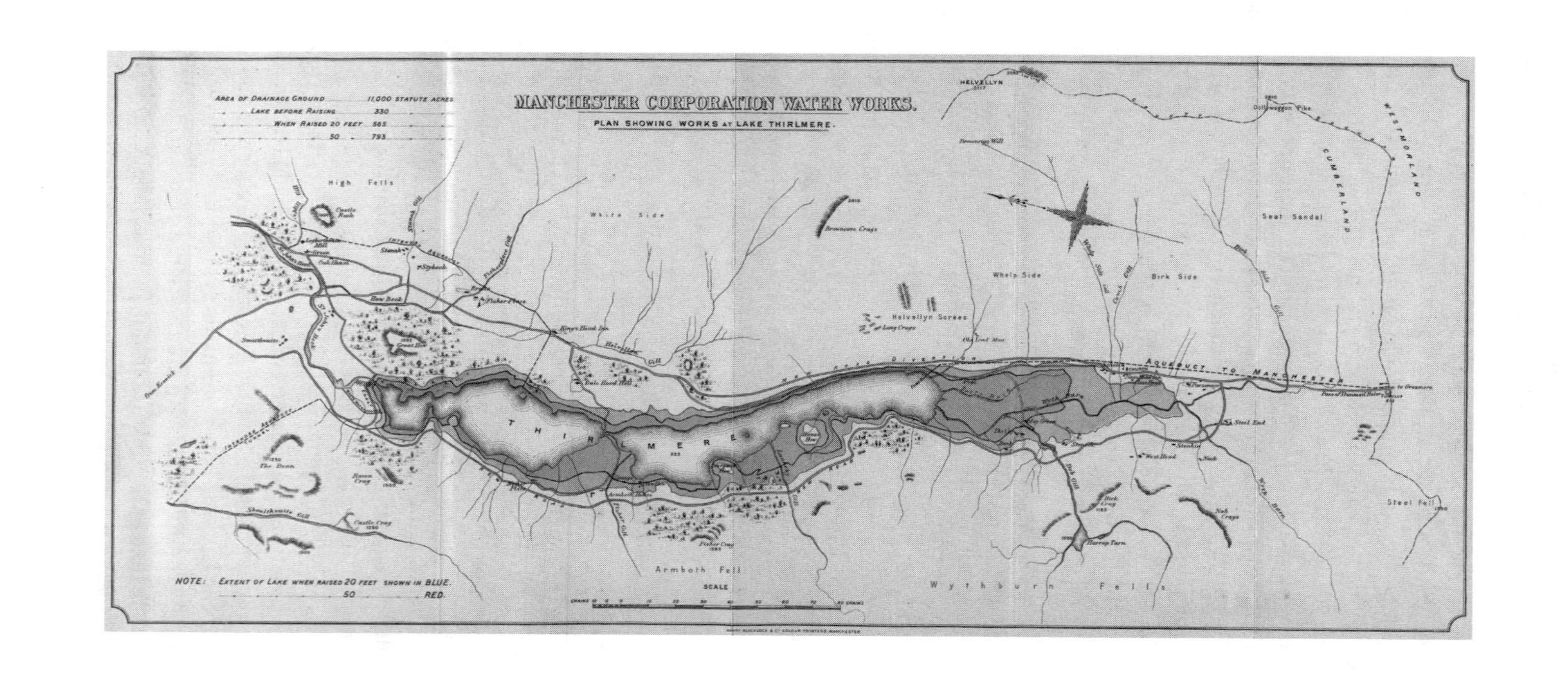
MANCHESTER CORPORATION WATER WORKS.
PLAN SHOWING WORKS AT LAKE THIRLMERE.
AREA OF DRAINAGE GROUND 11,000 STATUTE ACRES
LAKE BEFORE RAISING 330
WHEN RAISED 20 FEET 565
50 793
High Fells
White Side
Brownrigg Crags
Whelp Side
Birk Side
Seat Sandal
Helvellyn Screes
Long Crags
HELVELLYN
Dollywaggon Pike
WESTMORLAND
CUMBERLAND
AQUEDUCT TO MANCHESTER
Castle Rock
Stanah
Fisher Place
King's Head Inn
Helvellyn Gill
Dale Head Hall
Smaithwaite
THIRLMERE
Armboth Fell
Fisher Crag
Raven Crag
The Benn
Castle Crag
Shoulthwaite Gill
Harrop Tarn
Dob Gill
Steel End
West Head
Nab Crags
Steel Fell
Wythburn Fells
Pass of Dunmail Raise
to Grasmere
NOTE: EXTENT OF LAKE WHEN RAISED 20 FEET SHOWN IN BLUE.
50 RED.
SCALE

lance.[20] If Vane's negotiating position was not completely transparent at this point (that is, his right to make such an agreement was still being contested), neither was that of Manchester. Although the Waterworks Committee had already decided to build its new reservoir at Thirlmere, an internal memorandum reported that "when the preliminary interviews took place the design of the Corporation had not been published, and it was represented to Sir Henry Vane as the fact was that the adoption of Thirlmere depended upon his selling the Manor as otherwise the Corporation would go elsewhere and he would lose his opportunity." Perhaps because of these mutual misrepresentations, the initial negotiations proceeded so smoothly that "no attempt was made by the parties to reduce the terms of the arrangement to writing."[21] Both parties had subsequent cause to regret this amiable omission, as they spent the next year wrangling over divergent understandings of a few relatively minor provisions of the original bargain. Sir Henry was particularly reluctant to give up his right to hunt and shoot on the fells above Thirlmere; ultimately these rights were guaranteed for his lifetime, and in February 1878 his solicitor sent the final purchase documents to Manchester.[22]

By this time the Corporation's negotiations for the manor of Legburthwaite, which included farms and uplands along the northern shores of the lake, as well as the lake itself, were also far advanced. These negotiations were much more complex and emotionally fraught than were those for the manor of Wythburn. Unlike the Vanes, the Leathes family had actually lived at Thirlmere, in Dalehead Hall, the only lakeside residence that was even moderately imposing. One of the lake's alternative names—Leathes Water—testified to their long association with it (others included Wythburn Water, after the largest lakeside settlement, and Brackmere, which belied the Victorian encomia on its clarity). The first Leathes (then spelled Laythes) bought the manor in 1577, and the family then lived as unremarkable members of the local gentry for several centuries. They were notoriously jealous of their monopoly on the right to fish in the lake, having on one occasion successfully defended it against a challenge from the lord of the manor of Wythburn, first in court and then, when that victory proved not to be definitive, with weapons.[23] The estate was entailed, and so, when the direct male line failed in 1806, the manor passed to Thomas Stanger, a nephew, who adopted the surname of Leathes along with his inheritance. He lived in the south of England and mostly used Dalehead Hall as a summer residence. His son Thomas Leathes Stanger Leathes, on the other hand, grew very attached to the manor and the lake. He fiercely rejected the first Manchester overtures, but he died at the age of eighty-five in

1876, when the Thirlmere Scheme was just getting under way, leaving only a few loyal tenants to carry on his campaign of harassment and resistance.

Fortunately for Manchester's plans, his older son, George Stanger Leathes, who inherited the estate, saw things differently. He had lived in Australia for twenty years and was, in the shrewd estimation of the Manchester negotiators, more interested in money than in sentiment.[24] In the spring of 1877, his solicitor informed the Waterworks Committee that his client "might be inclined to sell the whole estate, but that the estate was entailed, and his brother, the next male heir, has . . . openly avowed his intention . . . to oppose most strongly the Waterworks scheme."[25] George had good reason to fear that his younger brother, Leonard Stanger Leathes, would take vigorous action; on previous occasions Leonard had sued to prevent him from cutting timber and opening mines, on the grounds that these profitable activities would spoil the beauty of the estate. The solicitor promised, however, that if an accommodation could be reached, George Stanger Leathes would "do all he could to support the scheme in Parliament."[26] The brothers thus engaged in a struggle that mixed self-interest with principle (at least in the case of Leonard), playing out as a family soap opera the aesthetic, ideological, and political conflicts that were also to occupy a larger national stage.

That is not to say that they kept their differences private. In November 1877 the brothers and their solicitors exchanged a series of hostile letters to the editor of the *Standard,* in which Leonard strongly associated himself with the growing opposition to the Thirlmere Scheme. He began by identifying with small local landowners, who were, he claimed, economically unable to resist Manchester's offers whether or not they favored the projected reservoir; only then did he turn to his personal stake in the case. He aggressively denied rumors that the lake and the estate had already been sold, on the grounds that such a sale was legally impossible, since "my brother is simply tenant for life . . . without power of selling. I am the next tenant for life, and my infant children are the heirs presumptive, and without our consent no sale can be effected." In resisting "this proposed act of Vandalism," he finally made common cause with "every man in . . . Cumberland and Westmoreland" and beyond, since Thirlmere was "the recreation ground not only of Manchester but of the whole of England." George's solicitor responded only to the legal assertion, pointing out that "under the Public Health Act, 1875 . . . a tenant for life can sell to an urban authority" for certain purposes, of which water supply was one, that both the estate and the lake had been sold the previous August, and that Leonard had been sent a copy of the contract. Leonard's solicitor pithily responded that "the alleged contract is not worth the paper it is written on." A letter to the *Times,* which responded to claims made by the Waterworks Committee, more elaborately distinguished between "lands" and "waters" for the purposes of the Public Health Act, and asserted that the "mansion, mines, and plantation" that formed part of the Leathes estate did not fall within its jurisdiction.[27]

Leonard maintained his opposition as long as he could. He filed a petition against the Manchester Corporation Bill of 1878, in which he emphasized the strength of his entailed claim on the property (his brother was fifty years old and had only daughters) as well as the charm of the lake: "among the loveliest in the Lake District."[28] He repeated these points in his testimony before the House of Commons Select Committee on the Manchester Corporation Water Bill, also praising the historical distinction of Dalehead Hall and lamenting that if the reservoir were to be constructed "as a county family, we are stamped out—blotted out."[29] His activity and outspokenness made him a target for the advocates of the Thirlmere Scheme, as well as for his brother's legal representatives. In his opening remarks before the select committee, Sir Edmund Beckett launched an ad hominem attack on all these claims to distinction and social standing by ungenerously describing Dalehead Hall as "a whitewashed house, with a small room, on each side of the front door, and a kitchen."[30] By easy inference, the occupant of such a dwelling could not pretend to the county status on which Leonard based his claims, nor was someone who so misrepresented his own position likely to be a reliable source of information about other matters. (A less partisan observer called Dalehead "a substantial, old-fashioned country house," furnished with "delicious old cabinets and wardrobes, massive sideboards, and chairs that would delight the heart of Mr. [William] Morris."[31])

Beckett used similar tactics against the Countess Ossalinsky, the owner of the only other lakeside property of any significance, whom he disparaged as "a foreign countess with some unpronounceable Polish name." He found Armboth House, her self-described "ancestral mansion," much less impressive even than Dalehead Hall (as, indeed, it was).[32] His characterization of the countess was less accurate; she belonged to a deeply rooted, albeit unremarkable, local family named Jackson, the exotic surname and title being the residue of a brief and unsatisfactory marriage in her youth. Her lakeshore property was relatively small—only 714 acres—but it was essential to Manchester's project, as were the five farms totaling an additional 850 acres that she also owned.[33] But if Beckett was vague about her origins, he was precise in his estimation of her motives. Although she had "declined to . . . consider the question of purchasing her land in any way" in November 1876, this refusal seems to have been part of a bargaining strategy, for a mere two weeks later she was "prepared to negociate for the sale of her estate at Armboth."[34] These negotiations stretched on for years. The countess recognized the strength of her bargaining position, which had been enhanced by the Corporation's acquisition of the two larger estates, and she intended to make the most of it. The Corporation initially offered her twenty-eight thousand pounds for her property, but she held out for arbitration and was ultimately awarded seventy thousand (plus compensation for her sheep). So unjust did the procedure leading to this settlement seem to the authorities governing Manchester and other large towns, that the Association of Municipal Corporations lobbied swiftly and successfully for a legislative revision of the arbitration process, which was enacted in 1883.[35]

Value for Money

The Countess Ossalinsky may have been the most persistent and troublesome of the owners and occupiers that Manchester had to buy out as the Thirlmere Scheme took shape, but she was far from unique in either her tactics or her goals. Although the manors of Wythburn and Legburthwaite accounted for much of the eleven-thousand-acre reservation surrounding the lake, the area also included a number of smaller landowners, with each of whom the Corporation had to come to a separate agreement. Most of these transactions were routine, but some cases required special attention. To persuade George Hynde to sell his farm, for example, Harwood promised that his widowed mother could live out her life in the farmhouse, and to reconcile the old woman to this change of landlord, he called on her frequently, each time bearing a packet of tea as a gift.[36] Often the transfer of title did not bring complete control of the property. As the experience of both the Vanes and the Leathes demonstrated, rights to fishing and hunting could be independently alienated, as could rights to cut timber and to mine metal or coal. More generally, landowners tended to lease their farms to tenants, often for extended terms. Such leases were not nullified by a change of ownership. Further, according to local custom, sheep belonged to the landowner and were only leased to the tenant, which meant that Manchester had to buy the sheep along with the farms—a double burden, since the Corporation had to pay extra for animals it did not want. Traditional arrangements also governed access to the uplands where the sheep foraged for most of the year. These unfenced areas were leased in units called stints, each of which was not a specified territory but the right to pasture ten sheep. As Harwood later reflected, "it will therefore be manifest that great difficulties had to be overcome in adjusting these various rights . . . in the purchase of all interests."[37]

In addition to acquiring every small property within the Thirlmere drainage area, and every lien on properties large and small, the Manchester Corporation had to pay landowners along the entire hundred-mile route of the cut-and-cover pipeline for the right to traverse their properties and for the continuing right to access the pipes for repair and maintenance. All this negotiation kept many lawyers busy for many years. The final arbitration concluded in December 1894, a few months after the official opening ceremonies for the reservoir and pipeline. It involved compensation to the spendthrift Earl of Lonsdale for several kinds of damage to an estate, just south of Thirlmere, through which the pipeline passed, as well as payment for stone excavated on his property in the course of construction and then used to make concrete.[38] Lonsdale was represented by the same solicitors who had achieved such a handsome settlement for the Countess Ossalinsky, but with much less impressive results. The arbitration umpire was unmoved by arguments that Manchester's need for the easement should make it more expensive, and he was persuaded by expert testimony that the pulverized stone was of poor quality and negligible value. He accepted the Corporation's characterization of its negotiating position: that Manchester was "quite willing . . . to pay the value of the wayleave and for any damage done" but would resist any "attempt to levy blackmail" upon its citizens. In the end the Lonsdale estate received only £1,991, much less than the £12,904 originally claimed.[39]

These negotiations were tedious, protracted, and unexpectedly expensive, but they were not surprising. By the 1870s Britain had had more than a century of experience with large public works projects, in the course of which both landscapes and townscapes had been repeatedly torn apart to accommodate new roads, bridges, canals, railways, and monumental buildings. Approbation of such enterprises was seldom unanimous, but the weightiest objections generally came from a few predictable sources. Like the Countess Ossalinsky, owners of property to be compulsorily purchased (that is, to be purchased via eminent domain) feared that they would not be paid enough for their land; like the Earl of Lonsdale, owners or tenants of nearby property, or of property otherwise impacted by the project, feared that they would not be adequately compensated for their inconvenience. On the other side of the transaction, ratepayers or shareholders (in the case of privately funded enterprises) feared that their money would be spent with insufficient care or excessive generosity.

This fiscally cautious perspective was forcefully represented in the internal Manchester debate that preceded the approval of the Thirlmere Scheme, and perhaps with greater vehemence for having been initially suppressed. The existence of broad support for the project, both within the city council and among ratepayers at large, had not made the Waterworks Committee eager to hear dissenting voices, or indeed, to open its decision making to discussion at any level. On the contrary, the committee had consistently failed to share specific information about the developing Scheme even with the city council as a whole, according to Aldermen Matthew Curtis and John King, both former mayors of Manchester who, to the great irritation of their civic colleagues, ultimately testified in opposition to the Scheme before the select committee. They complained that the general meeting of ratepayers and property owners required to authorize the parliamentary campaign of 1878 had been inadequately publicized, so that attendance was relatively sparse and the proposal was rubber-stamped after only five perfunctory minutes of discussion—a maneuver they characterized as "trying to hoodwink the Council and the public."[40] Since the enabling legislation failed to pass the first time it was introduced, a second general meeting had to be called in August 1878 to authorize additional parliamentary expenditure. On this occasion, no hoodwinking was allowed, and the proposal received extended and vigorous consideration.

Mayor Charles Grundy opened the meeting with a detailed rebuttal of the objections that Curtis and King had raised in speeches and in print. Their objections reflected the quantitative concerns of hardheaded businessmen, questioning (with some internal inconsistency) the need for additional water supply, the expense of getting it from Thirlmere, the adequacy of the new supply, the possibility of avoiding large cost overruns, and the extravagance of allowing other municipalities to share in the benefits of Manchester's investment, even if they paid for the privilege. Grundy's reply appealed to the members of the audience, who responded with enthusiastic "hear, hear"s and once with "laughter and cheers."[41] He was supported by the town clerk, Joseph Heron, whose arguments were less painstaking and more trenchant: "I cannot believe that Alderman Curtis honestly believes in what he has put before this meeting. I pay no compliment to his intelligence if I could imagine that he believes such trash as that will go down with the public of Manchester."[42] As was often the case, Heron had his fingers on the pulse of his constituency. Although Curtis and King, along with several others who opposed the Scheme, were allowed to have their extensive say, their listeners grew increasingly restive, ultimately interrupting them with hisses, groans, and demands for a vote. After a show of hands (but no counting), the mayor declared that the measure had been approved "by a large majority." A few diehard resisters nonetheless insisted on a formal poll of ratepayers, undeterred either by the expense of such a survey or by cries of "shame" from their fellow citizens.[43] The poll was duly carried out the following month, at a cost to the city of over eight hundred pounds; the unsurprising tally was 43,362 in favor of continuing to prosecute the Thirlmere Scheme and 3,530 against.[44]

Still, not everyone was persuaded. Alderman King, for one, continued to fulminate in opposition to the Scheme. Five years after the Manchester Corporation Water Act of 1879 received the royal assent he wondered in print "to what extent statements made at that time . . . have been verified," and he remained unshaken in his earlier conviction that Bateman and the Waterworks Committee had pulled the wool over the eyes of the city council and the population at large.[45] As before, he presented his challenges in terms of numbers—of money spent and water gained. By this time, of course, his criticisms could have had no practical effect. But even earlier, when they might have posed a threat, proponents of the Scheme found them easy to counter. A local consensus about the need for additional water had already been achieved. Moreover, the Manchester city fathers spoke the same language and shared the same underlying values. At the height of the controversy, King and Curtis broke ranks with their fellow councillors, acknowledging the aesthetic objections to the projected reservoir and even contributing financially to the organized opposition. (In hopeful if ineffectual response, the Thirlmere Defence Association had offered to "render them all the assistance we could."[46]) But fundamentally, their concerns were focused on the most literal (or at least the most reductive) kind of value, that represented by money. The Manchester Corporation understood such pragmatic objections and was well prepared to deal with them. It could draw on the expertise of engineers, scientists, and businessmen to provide abundant countervailing numbers of its own.

Unreal Estate

Values that were expressed in terms of thoughts or feelings presented a very different set of challenges, more difficult for the Corporation to confront. In his address to the public meeting of August 1878, Mayor Grundy marveled that "perhaps there is no bill, at any rate no bill not having the character of a Government measure, that ever exacted the large amount of public attention and public interest that this Thirlmere Water Bill has done."[47] He contrasted the uproar surrounding the Thirlmere Scheme with the more routine resistance that the city of Glasgow had encountered in its quest to convert Loch Katrine into a reservoir two decades earlier. The preliminary authorization procedure for that project had been even more time-consuming, requiring three parliamentary campaigns rather than the two that ultimately sufficed to authorize the Thirlmere Scheme. The Glasgow campaigns may also have been more costly. Comparison is difficult because, in addition to tens of thousands of pounds spent on accountable expenses, resistance to the final bill was eroded through the lavish indulgence of what a later chronicler characterized as "political graft, financial brigandage, capitalist greed and landlord gluttony."[48] That such pecuniary tactics achieved the desired end, however, indicates the conventionality of the objections to be overcome. Opposition to the Loch Katrina reservoir came from those who had something concrete to lose, from those (often the same parties) who had something concrete to gain from obstructing the project, and, perhaps, from those who did not adequately appreciate "the genius and skill of their engineer," John Bateman, and his plan—but, in any case, not from people whose claims on land and water were aesthetic and intangible.[49]

The most effective resistance to the Thirlmere Scheme came from exactly this quarter. Although they were joined, more or less wholeheartedly, by landowners whose property Manchester coveted and by a few Mancunians who felt the project was imprudent or unnecessary, the most vocal and energetic opponents of the Scheme had neither a legal right to be considered in determining Thirlmere's future nor a direct financial stake in the outcome of the debate. Their claim to consideration was instead their deep appreciation of and concern for the Cumbrian landscape. This claim they opposed to the equally intangible claim implicitly advanced in 1866 by the Royal Commission on Water Supply and subsequently by the Manchester Corporation: that the Lake District, with its surplus water and small population, belonged in some sense to the industrial cities and towns of Lancashire. The self-appointed defenders of Thirlmere, and of the Lake District in general, understood their beloved region as distinctive and distinct, not as part of the hinterland of Manchester. They valued it for its inherent aesthetic, historic, and even moral or inspirational qualities, not as a storehouse of commodities to be exploited for the benefit of remote city dwellers.

Despite its best efforts, the Waterworks Committee failed to keep its interest in Thirlmere secret for long. The projected reservoir was only the latest in a series of perceived challenges to the integrity of the Cumbrian landscape, and anxious local residents had developed sensitive antennae. In 1875, when the Manchester City Council made its commitment to acquire an additional water supply, these antennae were already quivering in response to the unrelated and ultimately unrealized threat to extend the railway north from its existing terminus at Windermere. Robert Somervell's efforts to resist that intrusion had attracted the attention of one of the Lake District's best-known residents, the aging and eccentric cultural critic John Ruskin, who contributed a preface to the pamphlet *A Protest against the Extension of Railways in the Lake District.* Along with Ruskin's characteristically elevated and excitable rhetoric—he grouped the projected rail extension with other products of "the frenzy of avarice . . . daily drowning our sailors, suffocating our miners, poisoning our children, and blasting the cultivable surface of England into a treeless waste of ashes"—Somervell reprinted sympathetic articles from magazines and newspapers.[50]

At the heart of the pamphlet was Somervell's "The State of the Question," which rehearsed arguments and positions that would frequently resurface in the Thirlmere controversy. He rejected "the imputation of selfishness . . . always an easy and popular mode of attack," asserting that he wished, not to prevent "the humbler classes of society" from visiting the Lake District, but to preserve it "for those—and they are found in every rank—who can enjoy its unsullied natural loveliness." He challenged the general assumption that the "laws of economy" should necessarily trump all other considerations. Conflating the aesthetic and the devotional, he claimed that Britain's "materially prosperous districts" were marked "by the darkening of the light of heaven, and the defilement and destruction of the beauty of the earth"; within this system of values, the extension of the railway would be a "wanton act of destruction."[51] But argumentative practice was not the only benefit of this preliminary essay in opposition. Somervell's proleptic efforts to thwart the railway provided him with valuable experience in organization. And, as a result of Ruskin's encouragement and assistance (most importantly, the loan of his name and his address book), Somervell had a ready list of three or four thousand people, some of them also widely known, who were prepared to give money and sign petitions to support the preservation of the Lake District.[52]

In the summer of 1877 the Manchester City Council formally voted to proceed with the Thirlmere Scheme. This public and official step served in the Lake District as a belated call to arms. Residents who had previously muttered and grumbled about rumored land purchases now gathered at an inn at Grasmere to contemplate action. More than sixty people attended the first meeting of what was to become the Thirlmere Defence Association (TDA), and they contributed nearly three thousand pounds to the cause.[53] The goal of most members of the TDA was to block the planned conversion of the lake into a reservoir. They did not, however, explain their mission in such absolute or transparent terms. Instead they suggested, in a brief if somewhat legalistic statement promulgated late in the autumn, that "the object they have in view . . . is not to induce the legislature positively to deny access to the Lake District as a source of water supply, under proper conditions, either now or at any future time, in case a real necessity can be proved," but merely "to shew that the conditions under which the present application is made, are in many respects objectionable."[54] In support of this ostensibly moderate objective they urged a revision of the ordinary legislative review procedure, to allow consideration of the proposed waterworks in the context of a broader range of potential benefits and drawbacks.

Since there was no chance of persuading the Manchester Corporation to alter the decision it had taken, the TDA directed its arguments to a larger audience in order to muster sympathy for its position and, at least equally important, financial support for the coming parliamentary confrontation. Its members felt that "if the facts are once fairly brought before the public, the joint action of a large and influential body of persons will easily be secured."[55] Somervell described himself as "much the youngest and poorest" of the group—and certainly he was the most junior member of its governing committee, which was otherwise populated by local notables—but he nevertheless emerged as its major publicist, both composing the pamphlets by means of which the TDA sought to carry its campaign forward and distributing them from his home.[56] The pamphlets were at once reasoned and engaged, implicitly demonstrating by both their methodical tone and their factual content that the TDA was neither elitist nor sentimental. They succinctly distilled many of the points that formed the bases of more elaborate subsequent arguments by opponents of the Scheme.

The first pamphlet, variously entitled *Water for Manchester from Thirlmere* and *The Manchester and Thirlmere Water Scheme: An Appeal to the Public on the Facts of the Case,* was available for the affordable charge of one penny—one indication of the scope of its intended audience. It "conceded that a sufficient supply of water for the inhabitants of large towns has . . . been recognised as a matter of such importance that considerations of taste and sentiment . . . must not be allowed to stand in the way," but also

THE

MANCHESTER AND THIRLMERE

SCHEME:

AN APPEAL

TO THE PUBLIC

ON THE FACTS OF THE CASE.

ONE PENNY.

MANCHESTER: JOHN HEYWOOD.
WINDERMERE: J. GARNETT.
LONDON: SIMPKIN, MARSHALL & CO.

noted that this recognition was not absolute, since "Parliament has wisely decided" that each water project must be assessed on its own merits.[57] It then challenged the statistical analysis on which Manchester's need for additional water was based, as well as the assertion that no alternative reservoir site was available outside the Lake District. The comparison to Loch Katrine—a similar reservoir, also designed by Bateman, which, advocates of the Scheme argued, had been constructed without significant damage to the surrounding countryside—was also examined and dismissed: "The level of Loch Katrine has been little, if at all, raised. The lake has *not* been converted into a reservoir, but simply used as one."[58] There was even an attempt, albeit ironical, to quantify the aesthetic value of the Lake District: "We have no *price-current* for the beauties with which God has clothed our world, but the immense sums which people, of all degrees and classes, spend every year, to procure . . . the enjoyment which these beauties . . . can afford, would be sufficient . . . to prove to the most unsentimental mind, that the best and rarest materials for such enjoyments are very far from worthless."[59] (The TDA persisted in this attempt to beat the Waterworks Committee on its own hardheaded monetary ground; a subsequent pamphlet argued that "even looking at the matter in a commercial light, it is obviously desirable that the resources of the District should be dealt with upon system, and not thrown piece meal to the first comers."[60]) The TDA's opening sally concluded on a pragmatic note, requesting that anyone interested in actively supporting its cause by signing or circulating petitions contact Robert Somervell.

The success of this appeal was evidenced in a more elaborately produced pamphlet, entitled *Thirlmere Defence Association. The Case Restated.* Even the penny edition included a colored map of the lake, showing its current shoreline, its proposed new shoreline, and the intermediate area that might be converted to unsightly and noxious mud in the summer when the reservoir would be low. For sixpence, sympathizers could buy a clothbound version, enhanced by two illustrations of the lake and a selection of supportive articles reprinted from the periodical press. It covered much the same argumentative ground as its predecessor, with a few adjustments in response to emerging details about the planned reservoir. It also incorporated a newly proactive suggestion for preserving "the very heart of the Lake and Mountain District." Citing the example of the United States government, which had recently designated the enormous territory that became Yellowstone National Park "entirely for purposes of public recreation and enjoyment," the TDA urged that "a Government, having an enlightened regard for the highest interests of the governed, might well take such a region under its special protection."[61] Although subordinated to the main argument of the pamphlet, this was a radical proposal, especially in view of the very different histories of landownership in England and on the western frontier of the United States. The concluding request for assistance grew longer and more detailed as the parliamentary struggle approached. It explained the desirability of petitions presented to individual members of Parliament representing nonlocal constituencies, as well as the multiple benefits of financial donations: "One thousand pounds, raised in sums of 5s. or 10s. each would not only be a substantial aid in itself, but would be invaluable as an expression of opinion."[62]

Although most TDA members lived at least part of the time in Cumberland and Westmorland, from its foundation the group also attracted distinguished adherents from much further afield. The academic elite was well represented; the membership list included the principal of Brasenose College, Oxford, and professors of fine art, civil law, astronomy, and modern history from Cambridge. The headmasters of Rugby, Uppingham, and the City of London School also subscribed. It was not surprising that the bishop of Carlisle, in whose diocese Thirlmere lay, endorsed the preservation effort, just as it was not surprising that the bishop of Manchester praised the planned reservoir as the fulfilment of divine intention. The future bishops of Liverpool and Salisbury supported the TDA, as did the Gothic Revival architect William Butterfield and the landscape painter Alfred William Hunt. A few names on the list of subscribers promised practical assistance as well as moral and financial support. Robert Hunter was a prominent legal authority on commons and public rights of way. And among the many causes championed by the social reformer Octavia Hill was the preservation of common open spaces and the public right of access to them, in both town and country. Hunter was among the founders of the Open Spaces Society in 1865, and both he and Hill continued their struggle to protect the Lake District landscape long after the fight for Thirlmere had been lost. Hill immediately put her experience to work for the TDA by organizing a meeting at her London house. The *Times* reported that its purpose was "to consider what practical steps should be taken with a view to render the resistance to the scheme as widespread and vigorous as possible." According to Somervell, who traveled to London to address the gathering, it fully accomplished this goal, energizing the campaign and helping to recruit moral and practical support on a national scale.[63]

Hill also introduced Somervell to Alfred William Hunt's wife Margaret, a novelist who served as a valuable connection to the periodical press, supplementing the access already provided by other supporters, both local and remote. In addition to aging lions John Ruskin and Thomas Carlyle, who cast long if fading shadows over the Victorian intellectual scene, the TDA included a number of frequent contributors to popular and specialized journals. For example, James Clifton Ward, who worked for the Geological Survey, was the author of the definitive scientific account of northern Cumbria.[64] Frederic William Henry Myers published widely on literary and classical topics, as well as on psychic research. James Spedding was a distinguished literary editor. H. D. Rawnsley, a prolific essayist and poet, had only recently moved to the Lake District to become vicar of an Ambleside parish, but he threw his formidable energies into the fight to preserve the landscape of his newly adopted home. As was the case with Hill and Hunter, for Rawnsley opposition to the Thirlmere Scheme turned out to be only the opening skirmish in a long campaign.

Point Counterpoint

Newspaper and magazine coverage of the TDA's crusade clearly demonstrated that public attitudes toward development and preservation had changed in the three decades since Wordsworth's quixotic attempt to block the rail line to Windermere. A great deal of ink was devoted to the issue. The *Times,* for example, opened its letters column to debate about the Scheme. It printed a long letter from the bishop of Carlisle, in which he characterized Thirlmere as "among the choicest of the English lakes . . . absolutely free from villas and all that is villanous . . . as wild as it was centuries ago," and then invoked the "shade of Wordsworth" to help prevent "this piece of Vandalism." Another letter from a local TDA member was crammed with geological and meteorological details about the Thirlmere watershed. When, provoked by the publicity accorded the views of "the Bishop of the Lakes," John Grave, the chairman of the Waterworks Committee, submitted an expansive "answer to some of the denunciations of our scheme which have been circulated in various ways," the *Times* also printed several ripostes, including one from Octavia Hill, who asked whether men such as Grave who "scorn the form of this lake as moulded by . . . long ages" are "the men to whom we are going to commit one of the loveliest lakes and valleys our England owns?"[65]

Most journalists seemed sympathetic to—or at least respectful of—the TDA's position. So widespread was favorable press attention that the TDA published another pamphlet, at the bargain price of one halfpenny, entitled *Extracts from the Leading Journals on the Manchester Water Scheme.* Most of these extracts combined appreciation of Thirlmere's unspoiled beauty with an attempt to redefine utility; many tacitly adopted the TDA's assumption of a kind of intangible ownership ungrounded in legal deeds. On the same day that it printed the bishop of Carlisle's letter, the *Times* editorially endorsed his appeal, noting that "some few spaces are still left us in England . . . where the weariness bred of daily toil can find the healthiest refreshment, and among those the hills and meres of Cumberland and Westmorland hold perhaps the highest place." The *Daily News* agreed that "there are other necessaries of a life worth living, besides water. Among them is beauty." The *Standard* more aggressively feared the advent of "a system under which whatever beautiful scenery, whatever healthful resorts are left in England, will be defaced and despoiled for the real or supposed advantage of those monster aggregations of houses miscalled cities." And the *Pall Mall Gazette* admonished the Manchester City Council, stating that even if their need for water was genuine, "they must not expect to get it in the cheapest market, to the disregard of all but commercial considerations."[66]

Newspapers in Cumberland and Westmorland were predictably eager to chronicle local resistance, which so annoyed Chairman Grave of the Waterworks Committee that he called on the editor of the *West Cumberland Times* in order "to point out to him the absurdity of the statements of his correspondents."[67] The TDA even received support from newspapers published in such northern industrial towns as Leeds and Newcastle, which had, like Manchester, experienced the difficulty of supplying water sufficient to the growing needs of industry and industrial workers. For example, after a brief acknowledgment of the fears of Manchester ratepayers who, like Aldermen King and Curtis, worried about "the enormous increase of their debt" due to the manipulations of "irresponsible persons," the *Yorkshire Post and Leeds Intelligencer* expressed its core objection to the Scheme in the language of the TDA: "When Manchester . . . proposes to destroy the natural beauty of the choicest spot in Great Britain—to bring vast engineering works and desolation . . . into the loveliest playground perhaps in Europe—then we are bound to withstand such encroachments."[68]

The TDA reprinted similar encouragement from general audience periodicals like the *Saturday Review,* the *Gentleman's Magazine,* and the *Spectator,* to the editors of which the planned reservoir sounded "very big and ugly and revolting, and we are not surprised that the Cumbrians do not want the Manchester people to paint their lilies for them."[69] Publications with more specialized concerns also joined the chorus. For example, the *Field,* whose readership enjoyed country sports, doubted that "an artificial dam, planted with shrubs, and dotted about with rock-work, could ever be regarded as a substitute for the wooded crags which lend such special charm to the scenery of Thirlmere." The *Medical Press and Circular* argued that "pure air [and] delightful scenery . . . are as necessary for the preservation of health as water supply," while the *Builder* characterized the claim that the Scheme "will not spoil the district" as "rubbish."[70] In *Fors Clavigera,* his series of monthly "letters to the workmen and labourers of Great Britain," Ruskin more earthily denounced Manchester's plan to drain "the little lake of Thirlmere . . . into its water closets." The longer he fulminated on this outrage the angrier he became, extravagantly claiming that although Manchester contained "many amiable persons . . . and much general intelligence," it could no longer produce anything good—even its cotton had declined in quality—and, more generally, was "cowardly in war, predatory in peace; and . . . plotting . . . to steal, and sell, for a profit, the waters of Thirlmere and the clouds of Helvellyn." As punishment for these lapses and transgressions, he proposed that "the town of Manchester, or at least the Corporation thereof, should be put at the bottom of the Lake of Thirlmere."[71]

The resources of ordinary prose (or even extraordinary prose, as in the case of Ruskin) were inadequate to express the strong feelings of some TDA supporters. *Punch* published a colloquy among its mascot Mr. Punch, the Lady of the Lake, and the ghosts of the Lake poets Wordsworth, Coleridge, and Southey. In heroic couplets the poem elaborated the standard contrast between "Engineers accursed" and "holy Meres," before apostrophizing the projectors:

> Nay, canny Cottonopolis, your plan,
> Though by your Bishop blest, Punch grieves to ban.
> Your water-wants you must supply elsewhere,
> And for your greed of cash, that's not our care.
> Nor will we spoil our lakes to sluice your City,
> Or please a Vandal Water-works Committee.

After this rousing peroration, Punch's shadowy interlocutors vanished, leaving him "resolute to plead the cause of Thirlmere v. Manchester." (*Punch* did not always attack from these literary heights; a later prose critique was more characteristically entitled "Boring for Water."[72]) Still more extravagant effusions appeared outside the constraints of periodical publication. *A Voice from the Ranks: An Original Poem on Thirlmere Lake* rehearsed all the major objections to the Scheme (and some of the minor ones) in ninety-five thumping tetrameter quatrains. Written by a poet who described himself (not entirely persuasively) as a "Cumbrian working man," it featured strong emotion, unself-conscious cliché, and unjustified optimism:

> Methinks, that if our Rulers saw,
> From this elysian sylvan hill,
> The mirrored Lakes, and scenes below,
> They'd never pass the Thirlmere Bill.[73]

"A Voice from the Ranks:"

AN

ORIGINAL POEM

ON

THIRLMERE LAKE,

COMPOSED BY

A CUMBRIAN WORKING MAN,

JOHN WILSON GRISDALE,

LAKE ROAD, KESWICK.

KESWICK:
BAKEWELL AND FITZJAMES,
PRINTERS.

Nov. 17, 1877.]

In Parliament, Session 1878.

Manchester Corporation Theatre Royal.

KING CRŒSUS;

OR,

HARLEQUIN TOWN CLERK IRWELL

AND

THE THIRLMERE LAKE.

(An Unacted Fairy Extravaganza.)

DEDICATED *WITHOUT THE LEAST PERMISSION IN THE WORLD*
TO ALL CORPORATIONS, LOCAL BOARDS, AND
PUBLIC BODIES IN ENGLAND,
AND THE GENERAL BUMBLEDOM OF EUROPE.

BY AN "ODDFELLOW."

LONDON:
SAMUEL FRENCH, PUBLISHER, 89 STRAND.
MANCHESTER:
JOHN HEYWOOD, PUBLISHER, 143 DEANSGATE.
1878.

Price 6d., Post Free.

Although the anonymous author of *King Croesus; or Harlequin Town Clerk Irwell and the Thirlmere Lake,* who identified himself only as "an Oddfellow," shared the general perspective of the "Cumbrian working man," he chose to present his views satirically. *King Croesus* is a closet drama in three brief acts, in which Titania, the fairy Queen of Thirlmere, learns with horror of Manchester's designs on her lake, then tricks and humiliates the members of a delegation from the city council when they arrive heavily laden with purses and picnic baskets. Titania and her immortal attendants speak in heroic couplets (as in *Punch,* the prosodic sign of virtue), while the mortals from Manchester express themselves in vulgar prose. The cast also includes a couple of non-Mancunian mortals, who blur both literary and geographical boundaries. Professor Ruskin is practically one of the fairies, feeling their pain and speaking in their couplets. An American tourist speaks no more elegantly than do the interlopers from Manchester, nor does his name—Mr. William W. Shakespeare Esau Bung—promise a delicate sensibility. But he nevertheless takes the side of the angels, truculently confronting the civic officials with an even more expansive figurative claim to the Lake District landscape than that promulgated by the TDA on behalf of the English nation: "I say this 'ere lake and those there mountains are as much mine as your'n; they are the natural wild scenes of my ancestors' home as of yours. . . . I don't understand our old country being spoilt for nothing. . . . I shall like to catch you trying to buy our Niagara."[74]

The Manchester City Council was slow to recognize the novel nature and strength of the opposition spearheaded by the TDA, and slower to acknowledge its legitimacy. As late as January 1878, just before the Manchester Corporation Water Bill was introduced in the House of Commons, Joseph Heron wrote to his fellow town clerk in Carlisle about the "unusual publicity" that the planned waterworks had received. He incredulously reported that "it appears possible, although I can hardly bring myself to believe that it is probable, that the opponents of this scheme may attempt to induce the House of Commons to reject the Bill." Heron may have been expressing more confidence than he felt, however, for he prudently went on to urge his colleague to ensure that the members from his district voted with Manchester.[75]

Even after the unlikely eventuality had transpired and the Thirlmere Scheme had suffered a parliamentary check, the Corporation continued to regard the TDA's opposition more as an annoyance than as a serious challenge. The mayor, town clerk, and city council still seemed more troubled by resistance within the city administration itself, such as that represented by the two rogue aldermen, than by any external objections or more nebulous public disapproval. This attitude was exemplified by Ben Brierley, a popular local author who also sat on the city council, in his account of a Lake District outing arranged by the Waterworks Committee "as a return for the loyalty displayed by the Council" in resisting the arguments of Aldermen Curtis and King. Whether he was describing the tedious journey to Cumberland or the tasty hot pot lunch the party had enjoyed at Dalehead Hall or the undignified exertions of his fellow councillors on the hillside paths, Brierley's usual tone was lighthearted and good-natured. But he occasionally interrupted his cozy reminiscences to ridicule the TDA and other "sentimentalists," expressing his amazement "that the Waterworks Committee had patience to argue the point with their opponents." When the party arrived at Thirlmere, he recorded their response to the notion that the landscape would be impaired by the projected reservoir: "Oh, the laughter!—oh, the yell of derision . . . , as we contemplated the mischief that was to be done to the grandeur . . . of Helvellyn!" He noted that the progressive residents of Keswick looked forward to doing business with the influx of visitors that reservoir construction would bring, in contrast to the "half dozen . . . old fogies" who thought "that the lake land was laid out by the Creator for their especial pleasure."[76]

The Thirlmere Scheme was a cause behind which all classes of Mancunians could rally. Even the Christmas pantomime at Prince's Theatre reflected this popular consensus in its characterization of the wicked uncle, the Baron de Boeuf, "who bribed the detectives, who introduced American beef, who opposed the Thirlmere Scheme, and committed other equally terrible villainies."[77] Brierley had begun his career as an ordinary factory worker, and he maintained his old connections through his writing, much of which was in Lancashire dialect.[78] In contrasting the economic interests of ordinary citizens with the rarefied pleasures of a self-constituted elite, Brierley echoed a point frequently made by advocates of the Scheme. In the autumn of 1877, for example, as the TDA was planning its attack, an article in the *Manchester Examiner and Times* confidently asserted that "in the immediate locality they [the Waterworks Committee] find no opposition." This assertion reflected an unrealistically narrow understanding of the possible sources of resistance, as was demonstrated by the further claim that "the owners of the property required have met them very favourably" (itself a rather optimistic view of the ongoing negotiations). But it also reflected willful blindness or denial, as did the simultaneous claim in the *Manchester Courier* that the "so-called 'Thirlmere Defence Association'" consisted "entirely of persons who reside out of the Thirlmere District." It characterized them as "people in whose mouths the cant of aestheticism is always to be found . . . the slaves of modern sham-artistic affections." The strictly residential part of this claim could have been disproved easily by consulting the printed lists of subscribers, complete with addresses, that the TDA repeatedly published. But the author of the article had been satisfied with a single authoritative source of information, relying, like the franker author of yet another simultaneous article in the *Manchester Guardian,* on the "mass of information . . . placed at our disposal through the kindness of the Waterworks Committee."[79] The Corporation could count on the understanding and support of local institutions, whether ecclesiastical or journalistic.

The Waterworks Committee had provided this material in order to inspire a journalistic riposte to the early efforts of the TDA. Soon after the articles appeared, in late October and early November of 1877, the committee collected and reprinted them in a pamphlet that superficially resembled the TDA's *Extracts from the Leading Journals on the Manchester Water Scheme.* There were, however, significant differences. The most obvious was that the TDA selections represented a wide swath of the national periodical press, while the Manchester anthology was drawn exclusively from local newspapers. That this constricted geographical focus was not entirely a matter of choice was shown by the Corporation's attempts at damage control in the *Times* and in the Lake District. But outside of Manchester, only the *Daily News* had editorialized in favor of the Thirlmere Scheme, arguing that "the claims of beauty and picturesqueness . . . should not be urged in a merely obstructive manner" and that "people who denounce the proposal of the Manchester Corporation, especially bishops and clergy, should be very careful to determine that their opposition is quite devoid of selfishness, and that they are not in truth opposing cleanliness, which is next to godliness."[80] Even that newspaper had changed its editorial tune by the following week, however, and in any case its initial sympathy had as much to do with London's water problems as with those of Manchester. Like the critical coverage that had appeared in the nonlocal press, this article appealed more forcefully to the heart than to the mind.

Emotion also figured in the Manchester reportage, but there it was swaddled in analysis and information. All the articles that the Waterworks Committee reprinted were similar in content and strategy (this was not surprising, since they shared the same official source). Designed primarily to consolidate local support, they explained at length the need for additional water both statistically and historically (that is, with reference to Longdendale), the process by which Thirlmere had been chosen as the new source of supply, and the financial and engineering details of Bateman's plan. They dealt with the main points raised by the TDA more perfunctorily, simultaneously disparaging both arguments and advocates. Apparently they were persuasive; at any rate, similar convictions were expressed by less sober journalists. For example, the *City Jackdaw,* which commented on the arts as well as current events, and which included the "Claws of the Week" among its regular features, declared its allegiance in the form of a cartoon. With the flamboyant new town hall in the background, the

drawing showed a city omnibus, marked "Thirlmere" on the side and "Parliament" on the front, top-heavy with city councillors. At the reins of the overladen vehicle Mayor Grundy cracked his whip. "Can Grundy drive them safe to Thirlmere?" asked a worried bystander; his companion assured him that "Joe [town clerk Joseph Heron] will get them through Parliament if anyone can."[81]

Thirlmere in the Dock

Despite Heron's dismissive skepticism about the TDA and the *City Jackdaw*'s confidence in him and Mayor Grundy, the Manchester Corporation Water Bill encountered serious difficulties in the parliamentary session of 1878. The corporation originally assumed that the approval process would run the cumbersome course that was routine for such legislation. It has recently been characterized as "a boon for the Parliamentary agents, barristers, and solicitors whose job it was to grease the Parliamentary wheels," and this was also the way it appeared to contemporaries, at least to those troubled by "the difficulties to be over come by a Corporation projecting a great work for the public benefit" and by "the costly and cumbrous nature of the present machinery for the passing of measures connected with the health and prosperity of our great urban communities."[82] All large municipal improvement projects required parliamentary approval in order to proceed with financing and with negotiations for compulsory purchase. Such applications were called "private bills," because they affected only particular localities (unless they affected London, in which case, given the size and importance of the metropolis, they might be considered both public and private).[83] There was nothing very unusual about the announcement of the intended application in the *London Gazette,* which described the ordinary purposes of such bills: "to empower the Corporation to purchase or acquire by compulsion or agreement, or take on lease any lands, houses or buildings . . . which they may require" and "to empower the Corporation to raise further moneys."[84] The same issue of the *Gazette* included similar announcements for a number of other projects—railways, gasworks, roads, and harbors, as well as waterworks. Manchester had successfully introduced many such bills in the past, dealing with sanitation and public safety, among other civic responsibilities; the waterworks at Longdendale alone had required nine.[85] The efforts of the TDA, however, ensured that this past did not turn out to be prologue. As the leading Manchester barrister ruefully pointed out in his opening address to the parliamentary select committee, "Without the extraordinary amount of commotion made by different people," the committee would merely have had to listen to a pro forma recital of engineering details.[86]

The Manchester Corporation Water Bill was introduced in the House of Commons early in 1878. At its second reading, on February 12, the occasion for general discussion by the whole House, one member representing a Lake District constituency immediately moved that it be rejected and another seconded the motion. Admitting that the bill did not quite fit the ordinary conditions for rejection at this stage—that "its object was of a novel character and contrary to public policy or general law"—they nevertheless argued that the issues raised by the Thirlmere Scheme were somewhat novel and too problematic to be handled by the ordinary procedure for approving private bills. In particular, they pointed to the scale of the project (that is, the area that would be affected in the Lake District and along the pipeline) and to the possibility that Manchester was planning to commodify its new water supply by selling the surplus to other towns. Rather than approving the bill, they urged that a special select committee or a royal commission be appointed "to consider the whole question of the water supply of these manufacturing districts, in order to see what recourse may be made to the Lakes, and under what conditions and limitations"—in effect, a northern version of the royal commission that had considered the London water supply a decade earlier.[87] A David-and-

Goliath theme was threaded through their remarks and those of the other members who spoke in support of their motion. The Waterworks Committee did not "anticipate that anybody could say 'No' to Manchester, and they believed that Parliament would be no exception to the rule"—but it was hoped that "the feeling which animates the British House of Commons" would rather incline its members "to take the part of the weaker side," especially since "this is very much a question of who has the longer purse." Bullying tactics, it was urged, should not be rewarded: "If the Corporation of Manchester has chosen in some measure to forestall the decision of Parliament by buying property at a considerable cost, they must take the consequences."[88]

The routine next step for a private bill that passed the second reading was consideration by a select committee, but it was objected that private bill committees were too narrow in their focus—"bound down by technical rules" and liable to hear only testimony from those with "sharply defined interests" in the proposed legislation.[89] Such committees were ordinarily small and composed of members whose own interests in the legislation might be analogously narrow. Despite the contrary arguments of the Manchester members and their supporters, the House ultimately voted, not to reject the Manchester Corporation Water Bill or to appoint a special commission, but to refer the bill to a specially constituted committee; four members of this hybrid panel would be appointed in the ordinary way, by the Committee of Selection, and five would be named by the House of Commons as a whole. *Lloyd's Weekly Newspaper* wryly characterized the committee as "particularly and peculiarly select."[90] Along with extended size came expanded scope; the Select Committee on the Manchester Corporation Water Bill could consider the public interest in such things as "the scenery of the Lake," as well as the representations of people and institutions with private interests in the Thirlmere Scheme. In particular it was instructed to report on whether Manchester actually needed additional water; if so, whether it was essential to supply this need from the Lake District; and if that was so, to what extent and under what conditions.[91]

Presided over by Lyon Playfair, an eminent chemist as well as a politician, the committee—which also included Ughtred Kay-Shuttleworth, the son of the sanitation reformer, and Charles Darwin's friend and neighbor John Lubbock—began its hearings early in March and submitted its report about a month later. The proceedings strongly resembled a legal trial, although there was, of course, no defendant—or, to put it another way, there were two defendants, the Manchester Corporation and the TDA. In fact, as the nature of the opposing legal teams suggested, there were more than two. The promoters of the Thirlmere Scheme were represented by a unified group of four barristers. The barristers retained by the TDA headed a more numerous but less coherent assemblage, which included independent counsel representing individual landowners, municipal corporations, and railway companies, among others. In this case there was no strength in numbers; on the contrary, they reflected the disparate agendas that would ultimately allow Manchester to divide and conquer.

Despite its limited resources, the TDA could draw on its own ranks for legal expertise. John Harward, who had attended the organizational meeting of the TDA, was the senior partner in the firm of Harwards, Shepherd and Mills, solicitors whose clients included the Great Western Railway, as well as the owner of "a mansion house situate in the midst of an estate of considerable extent and of great beauty, at Grasmere."[92] Somervell described him as "an oldish man with a long experience of Parliamentary practice, who told us what we could do, and ought to do."[93] Harward managed the TDA's case with a firm and occasionally prickly hand, as Somervell discovered when he asked Robert Hunter for legal advice as well as moral and financial support. As solicitor to the Commons Preservation Society, Hunter had acquired a great deal of potentially useful experience, and he urged that, instead of the barrister he disparaged as "Old Cripps" (Harward's choice), the TDA engage a lead barrister with greater personal gravitas and a stronger intellectual grasp of the aesthetic arguments against the Thirlmere Scheme. He suggested an extreme heavyweight—James Fitzjames Stephen, who had recently returned to England after distinguished legal service in India. Not only did Harward stick with Cripps, but he threatened to withdraw from the case entirely if his authority was compromised, and he was rude enough to Hunter to preclude any further advice from that quarter.[94]

Once securely in command, Harward and his colleagues prepared for the impending hearings with meticulous care, drawing up elaborate instructions for the cross-examination of the important Manchester witnesses. (Harward also appeared before the select committee as a witness, in opposition to the Thirlmere Scheme; as he explained in his testimony, his firm had dissolved its partnership for the occasion, to ensure that he had no formal professional interest in the TDA case.[95]) As in any legal contest, the intention was to impugn both the arguments and the motivations of the opposition. The memoranda supplemented specific suggestions about possible lines of questioning with background information to illuminate "the style and character of the man" to be examined.

The instructions concerning Waterworks Committee chairman John Grave thus emphasized that he was a native Cumbrian who knew the district well, and also that his "speculative" inclinations had occasionally led him into or close to trouble. That neither age nor prosperity had diminished these inclinations was shown by the fact that "legal proceedings are now pending in reference to a bonus which is said to have been secured by him in respect of the sale" of a company with which he had been involved. By implication, he might have led the Waterworks Committee on a similarly questionable course. His local roots and his material success, however achieved, were widely admired in his home region, where he was regarded as "a sort of enchanter who had waved his wand over the Lake and . . . as if by magic become on behalf of the Corporation the owner of all." The solicitors urged that Grave be queried about the genesis of a petition signed by Keswick tradesmen in support of the Scheme, on the grounds that they had been led to nurture unfounded hopes of profiting from the anticipated influx of construction workers. The most elaborate set of prospective questions were said to "emanate from Manchester," and, although their source was not further specified, they echoed the concerns that Aldermen King and Curtis had aired in other forums, with their focus on the secrecy of the decision-making process and the unreliability of the estimated costs and benefits of the Scheme.[96]

The instructions for examining the wily Joseph Heron similarly mingled content and character. They noted, for example, the town clerk's inappropriate reluctance to respond to the solicitors' legitimate preliminary requests for information about Manchester's finances and water usage. They asked, with regard to the Manchester proposals on a variety of compensation and damage issues, "Is not the advantage all on the side of the Corporation?"—at once an acknowledgment and a critique of his effectiveness in defending the city's interests.[97] The most extensive set of instructions was prepared for the engineer John Bateman, who would be the major witness regarding the design of the reservoir and pipeline. The majority of the proposed questions predictably concerned technical matters. Some attempted to undermine Bateman's professional prestige by unearthing previous failures, others to damage his moral authority by implying that he had evaluated his plan only on engineering grounds and not in terms of its financial or social impact. But some anticipated a purely political challenge. For example, if a propitious occasion arose—although "considerable discretion will be necessary in cross-examining on this point"—it was suggested that he be asked whether his plans for future expansions of the reservoir encompassed the inundation of the church and graveyard at Wythburn, which would allow them "to pin Mr. Bateman to the Parliamentary misdemeanor of not having disclosed his whole scheme."[98]

Heron, Grave, and Bateman, along with Mayor Grundy, were the most substantial witnesses in favor of the Thirlmere Scheme. Bateman's initial testimony took three days, and he was called back after other witnesses had testified to supply additional information. The four spoke in detail and at length about the city's increasing need for water, its repeated attempts to supply that need, and how the projected reservoir and pipeline would satisfy their present and future requirements. They also attempted to anticipate the objections that would be raised by the opposition, to which end they enlisted the support of a miscellaneous group of supplementary witnesses. Some of these were Lake District residents, mostly from humbler backgrounds than the TDA members, who viewed the Scheme as potentially beneficial. Thus the driver of the mail coach between Windermere and Keswick, who had been passing the lake for twenty-two years, testified that the proposed new roads would allow visitors a better view of the lakeside scenery, although under cross-examination he revealed that he seldom looked at the scenery himself, since "he mostly minds his horses."[99] A Keswick innkeeper testified that the Scheme was unlikely to decrease tourism to Thirlmere, since almost none existed. One local landowner (who also worked as an engineer in Manchester) asserted that, not only were the fears of his TDA neighbors ungrounded, but "tourists are already more of a nuisance . . . around Grasmere than would be a few [Manchester Corporation Water Works] employees."[100]

The Corporation also called experts in various fields to corroborate its technical assertions. Several engineers praised the safety and efficiency of Bateman's design, and several geologists asserted that the Thirlmere rocks were stable enough to support the additional stresses that the reservoir would produce. Henry Enfield Roscoe, the professor of chemistry at Owens College in Manchester, confirmed that Thirlmere's water was pure and of high quality.[101] A local surgeon hoped that the drainage of land surrounding the reservoir would improve the health of residents in the Thirlmere area, where, he asserted, fevers and consumption were prevalent. A landscape architect opined that the reservoir would enhance the scenic qualities of the valley of Thirlmere, an aesthetic judgment, it emerged, that he had formulated immediately upon being engaged by Bateman.[102] Frank Buckland, the national Inspector of Salmon Fisheries and the most flippant of the witnesses, guaranteed the suitability of the projected reservoir for trout and salmon, although questioning revealed that he had only glimpsed the lake once, during a February storm, and that he knew nothing about either the kinds or the numbers of its piscine population.[103] Sir James Bain, who was the Lord Provost of Glasgow as well as a landowner and magistrate in Cumberland, cited his city's happy experience with Loch Katrine as proof of the wisdom of such bold and ambitious municipal undertakings.[104]

Despite occasional lapses of information or tact on the part of the subsidiary witnesses, the case for the Thirlmere Scheme survived its initial presentation well. The opposing counsel followed the lines of questioning suggested in the preparatory memoranda, but their primary targets—Heron and Bateman—responded robustly. Neither their credibility nor their conviction was shaken. Edmund Beckett and his legal team were able to mold all the testimony, even that of their shakiest witnesses, into a unified and coherent argument. Their antagonists had to work with more refractory material. The witnesses opposing the Scheme were more numerous, but much less consistent in their perspectives and their goals (in this the witnesses resembled the barristers themselves). Whereas only one petition had been presented in favor of the bill (that of the Manchester Corporation), more than thirty were presented against it. These expressed a multitude of grievances, some of which were in conflict with the fundamental aims of most TDA members—"the picturesque people," as Beckett had snidely denominated them in his opening remarks[105]—and some of which were completely unrelated to them.

For example, powerful opposition to the bill came from urban and rural districts that had no general quarrel with the plan to extract water from the Lake District but were profoundly troubled by Manchester's aggressive appropriation of this valuable resource. After all, the Royal Commission on Water Supply had reserved the water of Cumbria for the "manufacturing populations in the north of England," and not for one particular city.[106] As the representative of Oldham put it, "This Bill gives Manchester an enormous monopoly of water. Thirlmere does not belong one whit more to Manchester than . . . to any other town in the neighbourhood." Figuring the Corporation as a predatory animal, he alleged that its wealth and power had allowed it to "pounce upon Thirlmere."[107] Similar sentiments were expressed on behalf of Bolton, Blackburn, Preston, and Wigan, as well as several rural districts through which the pipeline was projected to pass. Their representatives pointed out that modern "high farming" techniques, which required intensive use of manure and chemical fertilizers, had contaminated local water supplies, as manufacturing had contaminated urban rivers, so that they had as much need as Manchester for remote sources of pure water.[108] These claims correctly assumed that the Thirlmere Scheme would provide an aqueous bonanza far in excess of Manchester's current needs. The existence of such a surplus and the possibility of disposing of it profitably had been one of the arguments deployed by the Waterworks Committee to muster ratepayer support for its decision. While this commercial prospect played well at home, it intensified the alarm of neighboring municipalities, which feared Manchester's water gouging almost as much as they feared its water gluttony. Not only did they want to share in the supply, but they wanted to share on reasonable terms. In several cases their earlier overtures had been rejected; for example, the mayor of Wigan had been disingenuously informed that "the supply system would not extend" far enough to accommodate his city.[109] These witnesses' opposition to the bill was thus a last resort, a matter of strategy rather than principle.

Landowners whose property might be damaged by the construction of the reservoir and pipeline constituted another substantial source of opposition. Some, although by no means all, of such landowners belonged to the TDA, but in most cases, the alliance reflected convenience rather than commitment. As the organization had acknowledged in one of its fund-raising publications, "it is best to state frankly, that landowners whose estates are actually passed through by the works, have, with a few exceptions, little special interest in opposing this Bill, except so far as a deterioration of the scenery around them may affect the value of all property in the neighbourhood."[110] Indeed, the more radical claims of the TDA, which could have been interpreted as criticizing rural economic activity per se, would have had little appeal to large agricultural proprietors. Thus the testimony of the agents for the Earl of Bective's estates in Westmorland and Lancashire, which straddled the proposed route of the pipeline, focused on the potential for continual privacy violations by waterworks employees engaged in maintenance or repair. Although the cross-examining barrister characterized the suggestion that money might not adequately compensate for such intrusions as "sentimental," the agents spoke pragmatically rather than emotionally, their assertions buttressed by previous experience with similar liens for pipes and rails. They were also concerned about possible interference with existing drainage patterns as a result of the excavations, but showed little interest in the aesthetic implications of the Scheme. In the end both agents acknowledged, with some resignation but without much resistance, that accommodation would have to be made. As one of them put it in response to the question "You would consider that it might be a public necessity, which would override private interests?": "Yes, exactly."[111]

The pipeline was also projected to cross the Westmorland estate of

Stanley Hughes Le Fleming, whose representatives presented his objections in a very different register. He retained a separate counsel, Edward Pember, who described Rydal Mount as "one of the most beautiful [properties] in that part of the Lake District, and . . . one of the most ancient."[112] That William Wordsworth had lived there for almost forty years offered still more powerful evidence of its romantic attractions, as did the numerous poetic tourists who persisted in paying their respects at this domestic shrine. But where real property claims existed, it proved difficult to restrict the argument to the rarified level of aesthetics and heritage. The allegations of potential injury to the estate's intangible aesthetic value were paralleled by more concrete claims—incommensurables like disfigurement and disruption interwoven with contour lines and spoil heaps. Pember nevertheless insisted that attempts to quantify such considerations were inappropriate, despite "the growing determination of witnesses in these rooms to treat this as a matter of money." He insinuated that such determination might reflect significant differences between the sensibilities of "a country gentleman" and those of projectors backed by "the purses of a large corporation." As he warmed to his subject, he became more precise in his discriminations: the advocates of the Thirlmere Scheme were people without taste, unable to recognize "unfitness and vulgarity," and inclined to "sneer . . . at traditions which [Wordsworth] . . . has made sacred." In conclusion he quoted Wordsworth's own critique of Peter Bell, into whose heart "nature ne'er could find the way."[113] Testimony by Le Fleming's estate agent and several other witnesses familiar with Rydal Mount similarly stressed its aesthetic qualities, but the cross-examination persistently focused on details of cost and timing, lengths and depths. In the minds of Manchester's representatives, if not of those of Le Fleming, testimony was the preamble to negotiation.

As a witness, John Harward found it similarly difficult to control the interpretation of his evidence. Testifying not as a solicitor but as the owner of a small estate near Grasmere, he characterized the benefit he derived from his beautiful Lake District retreat as "incalculable," and his agent asserted that any diminution of this benefit "could not be paid for in money."[114] The pipeline was also his immediate concern, on the grounds of inconvenience (his house would be "uninhabitable" during construction) and safety (he feared flooding and rock slides as a result of the rearranged topography), as well as aesthetics (the pipeline would be "an ugly scar along the whole face of the property").[115] He conflated his apprehension of personal loss with the injuries that might be suffered by the nation as a whole, or at least that part of it that shared his appreciation of landscape and tradition.

The Lake District was "a little rich retreat within the reach of all people," and the geometrical line of the aqueduct would be a "serious drawback to the beauty of nature" along its entire Cumbrian length.[116] Harward repeatedly invoked the specter of Wordsworth, who had "immortalised" every tree and stone, especially in the neighborhood of Rydal, although he inadvertently acknowledged that these poetic associations had enhanced the value of local property.[117] As was the case with other proprietors, his cross-examination focused insistently on quantifiable loss and compensation, concluding with the question: "Supposing we were obliged to come, would you rather we took the whole?" to which Harward cagily answered, "I will tell you when you are obliged to come."[118]

It was only the witnesses who owned no Lake District property, or at least none that might be purchased by the Corporation or suffer damage for which the Corporation might offer financial compensation, who could put forward the TDA's case in its purest form, emphasizing the loss that would be sustained by the entire British polity if Thirlmere were to be irrevocably altered. For example, Sir Robert Farquhar, a resident of nearby Grasmere, testified that "the preservation of the Lake district is a matter of very great national importance" and compared the proposed Scheme's effect on "the beauty of Grasmere and Thirlmere" to the effect "if some modern cockney painter were to put a dash of colour over a magnificent picture . . . at the National Gallery."[119] As W. E. Forster, another local resident and also a member of Parliament, testified, "Something remarkably beautiful . . . may be taken away from the possession of England, and I do not think that any compensation to the owners of property will pay the British people for that."[120] These assertions were supported by strong convictions rather than solid fact. Aesthetic judgments were, then as now, matters of taste, and proponents of the Scheme found witnesses (mostly engineers) willing to predict that the well-disciplined redesigned lake would be more beautiful than the irregular original. In addition, claims that the planned reservoir would irrevocably alter a pristine landscape could be scientifically contradicted. The Pleistocene glaciations had scooped out the deep valleys in which the Cumbrian lakes lay, and in the immediate wake of the Ice Age, they had been filled with a lot more water. At Thirlmere, according to the director general of the Geological Survey, "the lake was considerably longer . . . beyond the upper end as it stands at present."[121] It thus became possible to understand the flooding of Thirlmere in its reduced, hydrologically deprived 1878 condition, as the restoration of the more robust Thirlmere of an earlier epoch.

Manchester's representatives did not change any opposing minds with such tactics, although they certainly produced confusion among the less intellectually agile of the lake's defenders. In any case, the interventions of the select committee members during the hearings showed that, like the promoters of the bill, they preferred to deal with the issue in the most concrete possible terms. Criticism that resisted quantification—whether in terms of pounds and shillings, or gallons of water, or tons of excavated rocks and dirt—was difficult for them to evaluate, in several senses. Without numbers, it was impossible to compare interests or assess damages. But aside from such questions of computation, the arguments of the TDA led the committee into unfamiliar territory. This deeper opposition to the Thirlmere Scheme rested on claims that were absolute, novel, and difficult to assess. It invoked a nebulous new sense of ownership—a sense that the citizens of a nation should have some say in the disposition of significant landscapes even if they held no formal title to the property in question. If the select committee members had sympathized with the opinions of the resisters, or if they had shared them, there was no obvious way to address their objection in the bill, which dealt exclusively with the concerns of individuals and corporate bodies having material interests at the proposed reservoir site or along the proposed pipeline. Indeed, it was far from clear that, from a legal point of view, the British people had any claim to be paid for their loss, or even any right (within the limited parliamentary context) to express an opinion about the disposition of property to which they held such insubstantial title. In his cross-examination of Forster, Beckett referred to the mere suggestion that any such right existed as "rather communistic." He returned to the attack in his final summation, dismissively claiming that "all this national sentiment . . . has been got up by what my learned friend calls the press" and, more particularly, that "Mr. Forster . . . seems to think that because he lives here, and for some reason or other does not want the lake interfered with, he is entitled to talk of national sentiment and British interests . . . I am not inclined to take Mr. Forster as Magister Elegantiarum."[122]

Whether or not they shared Beckett's disinclination, select committee members accepted his general arguments. The most heartfelt pleas presented by witnesses therefore had no discernable impact on the legislation that ultimately emerged. But if they failed to move the stony (or perhaps merely legalistic) hearts of legislators, they struck the most resonant chords in the wider public debate provoked by the Thirlmere Scheme. From the beginning the TDA had claimed that "the one mountain region in England is . . . the property of all Englishmen; any injury to that beauty in which its value consists is a greater and more irreparable loss to the nation as a whole, than it is to the landowners in the district."[123] Versions of this claim recurred in many comments on the parliamentary proceedings, and on the subsequent progress of the Thirlmere Scheme. Thus in the preface to *King Croesus,* the "Oddfellow" allied himself with the "large body who have held up . . . both hands against the invasion of commons and open spaces, and mountain and lake districts, by companies, corporations, or public bodies for economy's sake."[124] Neither the arguments of Edmund Beckett and his cohort, nor the conclusions of the select committee, had the power to make this "large body" disappear.

The End of the Affair?

The select committee did, however, have the power to shape the Manchester Corporation Water Bill. With regard to almost every issue, it decided in favor of the Thirlmere Scheme. In its report, which appeared on April 8, 1878, immediately after its final deliberations, the committee agreed that Manchester had a pressing need for additional water supplies and that no such supplies were available within the immediate neighborhood of the city. Dismissing alternatives suggested by the TDA as unworthy of "serious consideration," it agreed that the Lake District was the logical source and that Thirlmere was the lake most suitable for conversion into a reservoir. With a nod to the arguments of the TDA, it acknowledged that "the public at large has also an inheritance in the beautiful scenery of these mountains and lakes," but it was confident that provisions included in the new preamble to the bill would ensure that "beyond the temporary inconvenience and unsightliness caused by the construction of the works, there will be little or no permanent injury to the scenery."[125] The single important revision of the original bill was the stipulation that Manchester provide water to various industrial towns and rural districts along or near the route of the pipeline.

Although the revised bill easily passed the House of Commons, not everyone was pleased with the select committee's conclusions. Before it was considered by the House of Lords, a dissident who signed himself "Credat Judaeus" (in effect, although not literally, "tell it to the Marines") wrote "a short history of the case" for "the uninitiated to enable them to appreciate the Report of the Select Committtee at its true value."[126] In his estimation, this value was very small. He rehearsed the arguments of Aldermen King and Curtis about "the existence of a most lawless state of things at Manchester," which had allowed the Waterworks Committee to operate without accountability or supervision. He regretted that the select committee had not seriously considered alternative sources of water suggested by various opposition witnesses, and he derided the feebleness of the "amendments inserted by the Select Committee for the protection, as is fondly hoped, of the beauties of the scenery."[127] But his fundamental complaint was about the credulity of the select committee members, who had allowed themselves to be buffaloed by Beckett's "nursery lessons" and "vague abuse," and by Bateman's "semi-sarcastic assertion."[128] The TDA published a more measured critique, requesting support for a second challenge when the bill reached the House of Lords, on the grounds that the select committee had not fulfilled its responsibility to make "independent enquiries of their own" into the merits of alternatives to Thirlmere.[129]

The subsequent history of the bill was anticlimactic. The bill did fail in the House of Lords in 1878, but not on its merits or demerits; rather, it was claimed that the changes made as a result of the select committee's deliberations were significant enough to have required a procedural alteration.[130] Somervell and his fellow TDA members "were wild with joy at our victory," and Ruskin extravagantly characterized "the arrest of the Manchester Thirlmere plan by the House of Lords as the most hopeful political sign of the last ten years,—showing at last some perception by the English Lords of . . . what their power is, and their duty."[131] Their antagonists, however, appraised this essentially bureaucratic check more realistically. As the town clerk reported to the city council, "It is lamentable to think that, as the consequence of undue regard . . . to mere technicalities . . . , the time and money spent . . . in promoting this important Bill will to a large extent be thrown away . . . It is at any rate satisfactory to believe that if obliged to wait until the next session, the Corporation may . . . reasonably expect to receive the support of the Government and the House of Commons."[132] Despite an angry scrawl in red pencil across the top of this report—"another instance how the money of the People is frittered away"—Heron's prognostication proved correct. The Corporation's investment of time and treasure (over sixty thousand pounds) bore belated fruit.[133] When the bill was reintroduced in 1879, it passed both Houses easily.[134]

Only four petitions were submitted against the 1879 bill. The one from the TDA was a pale shadow of its former self, asserting in contradiction of the claims of critics like "Credat Judaeus" that the revised bill was "free from many of the more serious objections" that they had previously entertained. The group's remaining reservations concerned the effect of the proposed dam on two local streams and, more generally, members' fear "that the construction of the works in the manner proposed by the Bill would cause unnecessary injury and damage, and they humbly submit that

Thirlmere just before its transformation.

further provision ought to be made . . . for preserving as far as practicable the characteristic beauty of the Lake and Lake District."[135] To some extent, this moderation represented an acknowledgment of the handwriting on the wall; even the enthusiastic Somervell recognized that after 1878 the TDA could anticipate no further victories—at least no decisive ones. Opponents of the Scheme did, however, manage to amend the final bill in several ways, inserting clauses stipulating (among other things) that "all reasonable regard shall be had to the preservation as well for the public as for private owners of the beauty of the scenery," that the scenic disfigurations caused by construction should be disguised by "forest trees or underwood indigenous to the Lake District," and that existing recreational access to the fells should be maintained.[136]

Behind the scenes, the pragmatic Harward had been orchestrating a more concrete capitulation. In October 1878 a deputation from the Waterworks Committee reported appreciatively on a visit to his Grasmere estate and recommended that the Corporation make a rather generous offer to purchase it. Such a purchase would, they suggested, not only procure a desirable and attractive property, but also "an amicable arrangement would unquestionably save a large amount in Parliamentary costs and materially lessen the trouble and anxiety . . . [of] another severe Parliamentary contest." Harward would also "be able and . . . willing . . . to assist in the negotiations with . . . other Land owners with whom he has in the past opposition been intimately associated."[137] And that was the way it turned out. Encouraged by Harward's example, most (although not all) of the other recalcitrant landowners decided to settle with Manchester rather than make a quixotic last stand, and the Waterworks Committee officially recorded its "pleasure in referring to the loyal and satisfactory manner in which these gentlemen [Harward and the other solicitors in his firm] have assisted the Committee."[138] The Corporation also agreed to pay the TDA's outstanding expenses in return for the withdrawal of its petition.

This change of heart or of tactics was variously construed. A celebratory retrospective, published on the occasion of the reservoir's completion in 1894, echoed the view of the Waterworks Committee that Harward and his coadjutors had been "converted into sympathy with the scheme."[139] But in a valedictory communication to their supporters, the officers of the TDA offered a different perspective. They reiterated all their original objections to the Scheme and attributed the ultimate failure of their opposition to defects in the parliamentary process. Among the most important of these defects was the incapacity to consider one of the greatest moral and political strengths of the TDA's position: "disapproval on national grounds, felt equally by persons residing at a distance, and by the inhabitants of the District itself." Parliament was much better suited to addressing interests that were "merely local and proprietary." They characterized their muted objections to the 1879 Bill as a rearguard action, designed to "render the engineering works . . . as little injurious as might be, and mitigate the evil generally to the utmost possible extent." They thanked Harward (who had recently died, so perhaps this was a case of *nihil nisi bonum*) and his partners for their mostly unremunerated service to the cause. And they gave notice that they did not regard the struggle as over: "It will now be the duty of this Association to make arrangements for a watchful attention to the progress of the works, so as to insure their being carried out in conformity to the above mentioned provisions."[140] This proved, like many of the other assertions in the message, to be unduly optimistic. As Somervell ruefully noted, "What we could not foresee or provide against was the brutal determination of the Corporation . . . to write 'Manchester Corporation Waterworks' . . . over what had once been a lovely and secluded valley."[141] Nevertheless the defenders of the lake continued their rearguard struggle, under constantly shifting organizational banners. If they had lost this opening battle, the war was not yet over; the publicity accorded their campaign and the widespread sympathy inspired by their cause were to change the shape of future confrontations.

4

The Cup and the Lip

The Manchester Corporation Waterworks Act of 1879 represented a beginning more than an end. Manchester's legislative victory resulted from the forcible overpowering of the opposition, not the conquest of their hearts and minds. In addition, once authorized to proceed with the projected reservoir and pipeline, the Corporation still faced many practical obstacles. Negotiations for the purchase and leasing of property around Thirlmere and along the route of the aqueduct would, as noted in chapter 3, continue for years. This lack of closure did not preclude starting construction, any more than the absence of parliamentary sanction had prevented the premature commencement of land purchases. Nevertheless, the Thirlmere Scheme did not take material shape for several years. The economic depression that was already under way when Manchester brought its bills to Parliament made the Corporation reluctant to embark on the massive financial outlay that the new waterworks would require. Concomitantly, the slowdown in trade had meant reduced demand for manufactured goods and, consequently, reduced industrial demand for water.[1] As it happened, annual rainfall in the late 1870s and early 1880s was consistently above average, and sometimes well above average, so that water supplies from the Longdendale reservoirs were more than adequate to fulfill the city's immediate needs.[2]

Economics and meteorology thus combined to hearten local doubters, who continued to nurse forlorn hopes of derailing the Scheme. The indefatigable Alderman John King returned to the attack in 1884. Because five years had elapsed, during which the city's fiscal situation had worsened, while its water situation had apparently stabilized, he argued that his earlier predictions had been confirmed. The city council had made a mistake in uncritically accepting Bateman's calculations with regard to consumption and supply, and it was about to compound that mistake by committing the city and its ratepayers to extravagant and unnecessary expenditure. Instead he recommended "strict economy, . . . incurring no needless outlay involving interest and sinking funds, . . . obtaining the best price for what water is sold . . . , and being satisfied not to extend our limits of supply."[3]

Both King's words and the Corporation's inaction suggested that the need for additional water seemed less urgent than it had when the case had been made before the select committee. At least, it did not seem so urgent to the people who held the municipal purse strings. Appreciative Victorian chroniclers of the Thirlmere Scheme explained the delay in terms of sparse money and copious precipitation, if they considered explanations necessary at all. Thus one account, tellingly subtitled "A Compendious History of the Promotion, Progress, and Construction of the Great Thirlmere Waterworks Scheme," peremptorily dismissed the issue of timing: "It is unnecessary to here occupy space in describing the various preliminaries which were necessary after the Act was sanctioned by Parliament."[4]

The working-class population of Manchester may have seen things differently, however. The economic conditions that inspired frugality in the city fathers had also produced widespread and persistent unemployment among the urban poor. The construction of the reservoir and pipeline would provide opportunities for many skilled and, especially, unskilled laborers. According to a recent historian of Victorian Manchester, popular pressure exerted through formal channels (newly broadened by the extension of the franchise in 1884) as well as informal ones (the fear of demonstrations and riots) galvanized the Corporation into action, or at least into the semblance of action.[5] Coincidentally, 1884 was a drought year, and by autumn the Longdendale reservoirs held only one-sixth of their full storage capacity, "which caused serious inconvenience in the supply of water, both for domestic and trade purposes, and consequently much anxiety to the Committee." In January 1885 the Waterworks Committee, citing only considerations of water supply and of wayleaves (permission to lay the water pipes over private land), reported to the city council that "the time has arrived for active measures being taken in connection with the Thirlmere Works."[6]

One man's activity might still seem like another man's procrastination. The city's reputation and treasury were at stake, along with its water supply, and despite the incentives to haste, the committee proceeded with deliberation. Its first move was to request the council's approval of its choice as head engineer: George Henry Hill, whose forty-year association with Bateman as pupil, employee, and partner had made him deeply

Pipes and tunnel, ca. 1889.

familiar with Manchester's extant and prospective waterworks.[7] For the first several years, much of Hill's time went to negotiating the numerous contracts for materials and services required by such a large, complex enterprise—administrative activities that would have been more apparent to officials, lawyers, and accountants than to potential employees or consumers. The annual statements of the Waterworks Committee tracked his managerial triumphs and setbacks. Thus, in 1885 "the depressed state of the iron market" had made it possible to purchase pipes at a bargain rate. In the following year, less gratifyingly, the contractor for one of the major tunnels, who had already built temporary workshops, roads, and huts to accommodate laborers, suddenly "found it was not remunerative" and stopped work. As a result construction was halted while Hill made alternative arrangements. Mere delay was not the only possible consequence of such an episode, however, nor the most potentially damaging one. Despite the interruption, the committee's primary reaction was relief that time-consuming litigation had been avoided, along with "any complications that would prejudicially affect the interests of the Corporation."[8]

By 1887, more than a decade after the Thirlmere Scheme had become public, the Waterworks Committee could report good progress on the tunnels and the pipeline.[9] Construction went forward steadily, subject, inevitably, to environmental and technical obstacles. The rain and snow that made the Lake District such a promising location for a reservoir caused repeated problems. At times, the weather was so bad that all work on the surface had to halt, and only underground excavation could proceed. The earth itself occasionally proved less stable than had been hoped or predicted. After the pipes had been laid across the Fylde valley in Lancashire, for example, the ground shifted so radically on one slope that they had to be removed and replaced by new ones.[10] The engineers encountered their most serious difficulties in the crossing of the Troutbeck, to the east of Windermere, at a point where the stream rushed through a steep ravine. In deference to the wishes of the landowner, the original plan had stipulated that the pipes be concealed beneath the river. This plan guided construction for three dogged years, during which period the works were repeatedly swept away by floods. Finally, the subterranean transit was abandoned, and the pipeline instead crossed the Troutbeck on an iron bridge suspended fifty feet above the stream, anchored to the ground by concrete buttresses and secured to the rock by iron bands. The protracted and quixotic attempts to bury the pipeline, along with the lesser but still substantial challenges presented by the bridge, occupied so much time that this section of the pipeline was not completed until the summer of

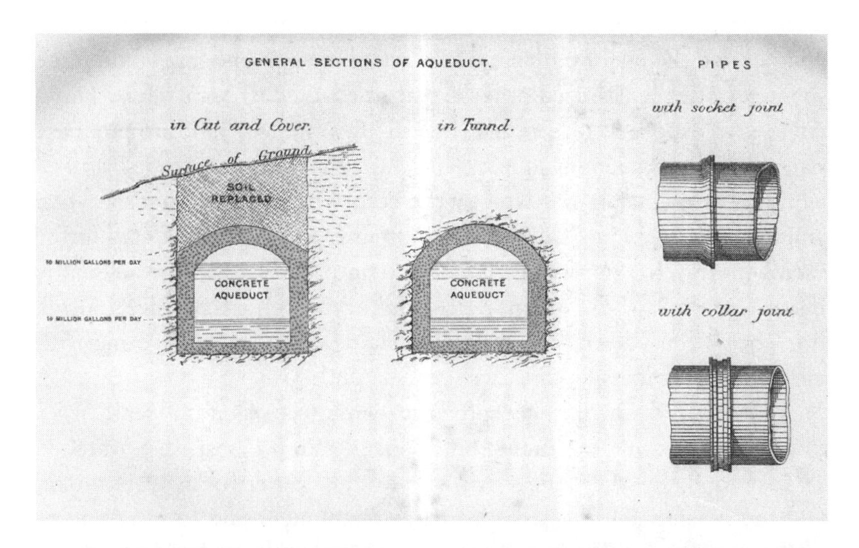

COMMENCEMENT OF THE THIRLMERE WORKS AT THE ENTRANCE OF DUNMAIL RAISE TUNNEL (3 MILES LONG). MEMBERS OF THE WATERWORKS COMMITTEE, ENGINEERS, AND CONTRACTORS, SEPTEMBER 24TH, 1886.

1893. It formed the last link in the long chain connecting Manchester and the Lake District.[11]

Work on the reservoir itself was by then well advanced. In the summer of 1890, the Waterworks Committee had invited the entire city council to visit the Lake District so that its members could "become acquainted with works which involved such a large cost, and with the difficulties that had to be encountered in their construction."[12] The committee had reason to be sensitive on the subject of expenditure. While King and his allies had queried the major outlays on real estate, equipment, and labor, more focused critics scrutinized other expenses charged to Manchester's ratepayers. An exposé of 1884 noted the use of committee funds to purchase nail scissors and a lemon squeezer, among other surprising items, as well as the generous salary received by the committee's doorman (thirty-two shillings per week) and the luxurious circumstances in which committee members traveled to inspect their works in progress.[13] Nevertheless the motive for the 1890 excursion was celebratory as well as defensive; indeed, it might also have been offensive, since Harwood was still answering criticism about its excessive cost three years later.[14]

About forty councillors accepted the committee's invitation and, in addition to inspecting the tunnels and pipelines in progress, they watched as Harwood, by then the chairman of the Waterworks Committee, laid the ceremonial first stone of the embankment and received an inscribed silver trowel and mallet, purchased by the members other than the chair "as a mark of their personal esteem and regard."[15] To justify the expense of this ceremony Harwood cited "the magnitude of the undertaking and the great responsibility the Council had in those works."[16] After the embankment had been completed, this earlier observance was itself commemorated with a large inscribed memorial stone. Beneath the seal of the city of Manchester, it states the bare facts, unadorned with rhetorical flowers: the date, the occasion, and the names of Harwood, mayor John Mark, the seventeen members of the Waterworks Committee, engineer George Hill, and the town clerk, William Henry Talbot, in that order. The letters of Harwood's name loom twice as large as those of Mark's; all the other names seem tiny in comparison.

Urbs in Ruris

The benefits of the Thirlmere Scheme began flowing south long before any water arrived. Hundreds of unskilled Mancunians were hired as navvies or construction workers (along with many itinerant laborers from Ireland), and spending associated with the Scheme pumped several million pounds into the pockets of Manchester merchants and manufacturers.[17] The impact on the Lake District was more complicated. Although Cumberland and Westmorland were accustomed to mass tourism, they possessed neither the physical nor the social infrastructure to accommodate a human influx of this scale and description. As many workers as possible were billeted at local farmhouses and other residences—even Dove Cottage near Grasmere, where Wordsworth had once lived, was used in this way.[18] For some people who lived in the immediate vicinity of Thirlmere, providing lodgings offered a final chance to earn money before they were evicted from their homes.

Most of the laborers, however, lived in specially built huts, at least after these temporary dwellings had been constructed. (Advertisements appeared in newspapers before construction began, and many hopeful navvies showed up before hiring had commenced and, therefore, before arrangements for their reception had been completed.)[19] The larger huts consisted primarily of a living room with tables and benches and a dormitory with ten or twelve double beds. Each hut was let to a married worker, whose wife and children cooked for the single men and charged them for their board.[20] Such accommodation was basic, although it afforded minimal domestic privacy for the managerial family. When describing the provisions made for the "comfort" of their laborers, the Waterworks Committee therefore preferred to emphasize its supplementary contributions. For example, a schoolroom had been built after "the children of the navvies and workmen were found to be wandering about the mountains, and it was manifest that a deplorable condition of things would soon result unless something was done to employ their time." When it was not needed for education, it served as a reading room and a chapel. And when harsh winter weather suspended all labor, and therefore all wages, it served as a soup kitchen, where "daily meals were freely given to all." According to Harwood, it was "impossible to describe the gratitude evinced by the partakers of the bounty to the Committee." The committee also established a temporary hospital, the location of which shifted as the works progressed.[21]

Harwood credited these benevolent interventions with the generally good behavior of the navvies, which he was pleased to emphasize because concerns about increased public disorder and crime had featured prominently in local resistance to the Thirlmere Scheme from the beginning. In his speech before the House of Commons Select Committee, for example, the representative of Leonard Stanger Leathes had predicted that during the long period of construction "a large number of workpeople presumably not of very elevated manners or pleasant dress, would saunter about the lake after their work is done." The rector of Grasmere may have been more realistic in anticipating their likely after-hours destination when he asked the Waterworks Committee to provide "a large, roomy, bright

Coffee House" for those workingmen who might prefer it to a pub.[22] The Waterworks Committee responded to these worries by subsidizing local constabularies, as well as by providing alternative recreational venues.[23]

Neither of these measures sufficed to prevent occasionally egregious drunkenness and brawling, such as the armed battle involving several hundred navvies, English against Irish, that began in a pub near Kendal and continued for several days in September 1890.[24] At least in part, these outbursts represented a response to dreary living conditions and difficult dangerous work. (Official accounts of the Scheme seldom recognized the hardships endured by laborers. H. D. Rawnsley was unusual when, in the sonnets he composed for the reservoir's opening ceremonies, he commemorated "the few unknown" who died during construction.)[25] Although it was acknowledged that "all sorts and conditions of men were represented among the navvies," including self-educated naturalists and quiet readers, most descriptions of them emphasized their affinity for "drink or crime." In 1890 a chronicler of the ongoing changes caused by construction blamed the navvies housed near Wythburn (ironically denominated "pioneers of civilization") for harassing churchgoers, schoolchildren, and more respectable drinkers at the Nag's Head Inn.[26] Thus Harwood's evaluation of their behavior, despite being buttressed by the testimony of the chief constable of Cumberland and Westmorland that the workmen "had conducted themselves in a most exemplary manner," was not universally shared. Even when the reservoir had been completed and the workers had gone, the memory of "the blatant navvies who made our peaceful valleys ring with strange oaths" was slow to fade.[27]

Equally persistent were memories of physical disruption. If the rough-looking laborers were sometimes cited as both aesthetically and behaviorally problematic, the results of their labor seemed even more disturbing. Local residents were shocked by the size and apparent disorder of the temporary encampments, such as the one that covered White Moss Common (near Rydal and the estate of Stanley Le Fleming) with huts and offices.[28] Much more striking, especially as work on the reservoir intensified, was what a travel guide published in 1891 characterized as "the defacement arising from chimneys and spoil-heaps" and "a smoke nuisance . . . [that] would certainly not be excused in Manchester itself."[29] Reginald Blunt, who toured the construction at about the same time, described the approach to Thirlmere from the south as a transition from the autumnal Lake District landscape of romantic convention—"the glory of colour on leaf and bracken and hillside, . . . Helvellyn . . . majestic, clad in a dazzling mantle of snow"—to "Manchester country," indicated symbolically

Setting in Foundations. (No. 1 Embankment.)

by "the mystic letters 'M.C.W.W.'" and atmospherically by "the workings of Cottonopolis in a film of dark grey haze . . . the grimy efflux from a score of furnace chimneys."[30] He used the language of desecration and defilement to evoke the radical transformation of the scene: "black huts . . . risen like mushrooms," "a huge quarry eating its wicked way into the side of Great How," the "dishonoured bareness" of an abandoned road, a well that seemed like a "bottomless pit," and an excavated hole that seemed like an "icy inferno."[31]

Although "the invasion of the navvy and the engineer" was visually (and sometimes aurally and olfactorily) overwhelming, it was only the emblem of permanent change, not the change itself.[32] The human tide receded with the completion of the reservoir, and the chaotic evidence of construction—"huge heaps of stone, sand, and cement, of black tarred sheds, snorting engines puffing dirty smoke, and barrels, cans, empty cases, straw and litter"—vanished into the reconstituted landscape. By 1900 a local observer could find, even around Wythburn, at the southern end of Thirlmere and therefore at the beginning of the pipeline, "hardly any sign . . . of the ugly upheaval and disorder which made portions of the valley look so unlike themselves."[33] He attributed this happy obliteration less to the perfunctory efforts of the Waterworks Committee—"soil was strewn over heaps of waste, and grass seed sown upon it"—than to the restorative powers of nature. In a mere six years "sun and rain, wind and frost combined to wear away crudities, . . . to wash away stains," aided by the slow, persistent rooting of moss, heather, ferns, and other small, hardy plants.[34]

The restoration of order did not, of course, mean the restoration of the old order. Within its ring of hills, the lake had been completely altered, in appearance as in function. The reconstituted valley could no longer have supported the people who previously lived and worked there, even if they had been allowed to reoccupy it. The Corporation closed the mines as soon as possible, although the operators of the Wythburn Mine struggled briefly against this decision.[35] The raised water level of the lake had submerged the old road along the eastern shore, along with many farmhouses and cottages, and almost all the lakeside buildings that remained above water were intentionally destroyed.[36] The hill sheep were evicted along with their shepherds. Despite the "old world character of the buildings" and its "atmosphere of remote antiquity," the hamlet of Wythburn was doomed to be "swamped out of existence" when the water level in the reservoir was raised to its full height.[37] Its small church survived, along with, to Rawnsley's great relief, one of its taverns. In 1916 he shared his persistent "fears about the possible removal of the Nag's Head Inn," which

Thirlmere, before and after the Scheme.

he characterized as a "famous hostelry," as well as his pleasure that Manchester had decided to spare it as long as its sewage did not "contaminate [the city's] mountain cup of crystal clear."[38] Dalehead Hall remained—apparently little changed, but transfigured through recontextualization. Still the seat of the (now corporate) lord of the manor of Legburthwaite, it had also become the remote headquarters of the Waterworks Committee.

Say Not the Struggle Naught Availeth

Once Parliament had approved the Scheme, the depopulation of the valley, along with its de-ovification, became inevitable. Manchester's plans had always required the elimination of competing economic activity within the Thirlmere drainage area, along with the expansion of the lake itself. But other changes were far from inevitable. Indeed some of them, at least in the view of the Scheme's most earnest opponents, had been expressly prohibited by Parliament. Although the Thirlmere Defence Association ceased to exist soon after its cause had been definitively lost, its spirit lived metempsychotically on. Defeat had neither changed the minds of its adherents nor dampened their commitment to action; and Thirlmere was not the only front on which they battled. In 1882 and 1883 their continuing engagement sparked an effective ad hoc campaign of resistance against a proposed railway line from Braithwaite, near Keswick, to the slate quarries near Buttermere, in the scenic heart of the Lake District—an achievement that the popular weekly journal *All the Year Round* celebrated as "great work."[39] But, as H. D. Rawnsley argued in a speech he presented before meetings of the Cumberland Association at Ambleside and the Wordsworth Society at Westminster, this small triumph offered

H. D. Rawnsley.

no cause for complacency. He reminded these disparate audiences that "if the Honister Pass steam-dragon has been baffled, other objectionable threats of invasion of Northern England's recreation . . . grounds are being made." The sources of danger included thoughtless local tourism boosters, such as the Lake District Association ("Heaven save us from such custodians of the Lakes!"), as well as greedy entrepreneurs from outside.

Rawnsley urged the formation of a permanent society, buttressed by strong national and international support, as "our only chance of keeping Lakeland inviolate." This society would work in concert with sympathetic members of Parliament, and with like-minded organizations such as the Commons Preservation Society and the Ruskin Society. It would (he hoped) be able to command "a considerable sum of money," and it would enjoy the support of "educated Americans [who] feel more keenly than we do, by a sadder experience, how scenery can be despoiled."[40] (At about the same time, for example, defenders of Niagara Falls were engaged in an analogous struggle against that site's "dishonor . . . disfigurement and desecration" by the forces of commercial and industrial exploitation.[41]) The Lake District Defence Society was established soon afterward, both its aims and its rhetoric recalling those of the TDA. A fund-raising pamphlet published early in 1885 announced that it had been formed "to offer a powerful and consolidated opposition to the introduction of unnecessary railways into the Lake District, and to all other speculative schemes which may appear likely to impair its beauty or destroy its present character."[42]

At that time the society had about two hundred members, a number that had nearly tripled by the end of the decade. Rawnsley identified Charles Eliot Norton, a Harvard professor and a distinguished man of letters, as eager to galvanize American support. Many of the society's subscribers had similarly elite backgrounds and similarly (if not equally) remote residences; the list was heavy with academics, clergymen, professionals, and people from the world of arts and letters, along with affluent merchants and manufacturers. Fewer than 10 percent of the members resided in the Lake District; the largest contingents (each about one-quarter of the total) came from the vicinities of London and of Manchester.[43] As Rawnsley tirelessly reminded his public, the threats to the integrity of the Lake District were constant and varied, and vigilance was required throughout the entire region. Nevertheless Manchester's implementation of the Thirlmere Scheme remained a special concern. In addition to the transient upheavals and disfigurements associated with construction, therefore, preservationists monitored alterations that promised greater permanence.

COMMON PIKE.

GRAYLING AND CHARR ($\frac{1}{6}$ nat. size).

Possible inhabitants of Thirlmere.

Every aspect of the lake and its immediate surroundings was perceived to be at risk, even when it was difficult to specify exactly what needed protection. For example, the parliamentary hearings had elicited widely differing predictions about the likely effect of the proposed conversion on Thirlmere's piscine population, with opponents prophesying degradation and advocates enhancement. These predictions had been based on equally divergent accounts of the number and nature of Thirlmere's existing fish. TDA member Robert Farquhar had extolled the "fine stock of lake trout, specimens of which are often caught in the adjoining becks," while the enterprise-oriented Frank Buckland, Inspector of Salmon Fisheries for England and Wales, had dismissed the lake's stock as "comparative rubbish."[44] Farquhar credited the Leathes' vigilant enforcement of their fishing rights with the alleged plethora, while Buckland blamed poachers for the alleged dearth. (Later naturalists suggested that the same lack of organic matter that made Thirlmere's water so attractively pure for the purposes of Manchester also made the lake a relatively poor habitat for fish.[45]) Both assertions had long historical pedigrees; the centuries-old restrictions on fishing meant that such speculation had seldom been empirically constrained. A survey of the lake counties published in 1777 had assured travelers that Brackmeer (one of several alternative names for Thirlmere) was "well furnished with pike, perch, and eels," while a similar account published ten years later warned that "Thyrillmere . . . is of but small value for fish."[46] This mixture of authority and inconsistency persisted as Lake District tourism increased and travel literature proliferated. Mid-Victorian anglers could thus read both that Thirlmere was "well stocked with trout, but the public are not allowed in boats upon it," and that despite vigorous preservation by the Leathes, the Thirlmere trout population had been decimated by the pike ("that most voracious of fishes") with which they shared the lake.[47]

The transfer of ownership did not make things much clearer. In his testimony before the select committee, the Manchester town clerk, Joseph Heron, had promised to "allow more boats upon [the lake] than Mr. Leathes has allowed."[48] Nevertheless (although their motivation different—to protect the lake's pure water from contamination, rather than to preserve all the sport for themselves), the Waterworks Committee members maintained the restrictive practices of Thirlmere's previous owners, mandating in 1878, even before the Scheme had received parliamentary approval, "that no person be allowed to use Boats on, or to fish in Lake Thirlmere" without their express permission.[49] That such permission might occasionally be given was suggested in the spring of 1879, when it was reported to the committee that six thousand Geneva trout and thirty silver char, provided by Buckland, had been transported from London to the lake. While Buckland did hope, in introducing nonindigenous food fish, to provide an "important angling ground" for "the citizens of Manchester, and . . . numerous tourists," he viewed the pleasure of sportsmen as merely a side benefit. His official report for 1878 noted that "Thirlmere being converted into a reservoir will afford an unexpected opportunity of forming a . . . hitherto unutilized lake into a fish farm. . . . The Mayor and Corporation of Manchester propose to cultivate these waters with non-migratory Salmonidae," including species imported from the south of England, Switzerland, and the United States.[50] Buckland apparently overestimated the extent to which the Corporation shared his particular aspirations. However strong its sympathy with the commercial and utilitarian spirit of his plan, Thirlmere never became an industrial-scale producer of fish, and it remained off limits to ordinary fishermen.

As a result, although the progress and eventual completion of construction focused increasing attention on the lake, reports about its aquatic fauna remained hypothetical. Even expert anglers writing for their fellow enthusiasts had to rely on inference and (often recycled) rumor. Thus in 1884, when the works were beginning, George Foster Braithwaite announced that all unspoiled Cumbrian lakes were "favourable to the growth of the salmonidae [salmon, trout, and char]" and that the nature of the fish varied according to the geological substrate of each individual lake; he therefore grouped the putative Thirlmere trout with the more accessible denizens of Windermere, Grasmere, and Buttermere. In 1899, after the reservoir was finished, the Lake District volume of the "Angler's Library" series wistfully announced that "Thirlmere is closed to the angler; although . . . it has been restocked with trout. The probabilities are, from what can be observed from its shores, that the lake teems with good trout." A decade later a guidebook to the Lake District speculated that "in the course of time, *angling* in Thirlmere may be expected to be very good," basing its sanguine prognostication on (the by then long-deceased) Buckland's high opinion of the Geneva trout he had introduced forty years earlier. And even in 1922 a fishing guidebook hypothetically claimed that "the possibility now is that it simply swarms with big fish" although, due to official vigilance, "it was easier to get to Paradise than to Thirlmere."[51]

Some inferences could be based on observation of bodies of water that were near the converted reservoir or that had similar origins and functions. Local streams provided circumstantial evidence about the effect of the dam that was, at best, only moderately encouraging. Thus, Helvellyn Ghyll offered anglers "some of the most remarkable and impressive scenes that Cumbrian mountains can afford," but its trout were neither "large nor plentiful." St. John's Beck "used to be a capital trouting stream, but owing to the abstraction of water from Thirlmere by the Manchester Corporation, and to the trout not having free access to and from the lake, the trout fishing has somewhat deteriorated, though still good."[52] In contrast, the reservoirs constructed elsewhere by large industrial towns provided abundant—and even exemplary—fishing, although, since they were managed according to more accommodating principles, their bounty may have suggested Thirlmere's potential rather than its actual condition. Loch Katrine nurtured plentiful salmon, trout, and char, and the Glasgow Corporation exacted no fee for catching them; fishermen could easily rent boats either from lochside farmers or from nearby hotels. The Liverpool Corporation had always planned to exploit the sporting possibilities of Lake Vyrnwy, which was created by flooding a remote and dramatic Welsh valley. The reservoir began to produce "good trout" soon after its completion in 1889, and not long afterward an aficionado pronounced that "for trout-fishing with the fly there is no better llyn [lake] in Wales."[53]

Somewhat easier to corroborate or dismiss were fears about Thirlmere's terrestrial surroundings. Among the many aesthetic issues raised when the Thirlmere Scheme was initially proposed was the fear that the shoreline of the expanded lake would be an eyesore, especially at times when the water level of the reservoir was unusually low due to heavy usage or drought. Thus the instructions for cross-examining Bateman before the select committee urged that he be confronted with "direct evidence of the strongest character not only that the lake does deposit a coating of mud where the water is still but that . . . when the coating is exposed it is both unsightly and offensive."[54] In response to a similar private inquiry, Thomas McKenny Hughes, the Woodwardian Professor of Geology at Cambridge, received official assurance, firm although not exactly to the point, that "a number of years must elapse before the Lake could be drawn down to its present level, after the Embankment has been made and the water supplied to Manchester."[55] Although, after the reservoir was completed, George Henry Hill assured a meeting of his fellow civil engineers "in consequence of the country being particularly rocky and disintegrated" there would not be "any objectionable appearance" even when "the water was withdrawn to any great extent," these fears were subsequently shown to have been very reasonable.[56] Low water levels did indeed expose tracts of mud, and the motion of water and wind produced continuing erosion along large sections of the shoreline. But even this cloud could be perceived to have a silver lining, transforming the product of human engineering into a product of the same forces that had shaped the other lakes for thousands of years. According to a geologist who examined Thirlmere's shoreline several years later, its unstable outline and shifting sands explained why, "to the ordinary tourist . . . there are only one or two places where the artificial nature of the [lake] . . . is obvious." And—an unanticipated further benefit—the shifting sands provided scientists with "an opportunity of observing and recording the effects of the forces of nature."[57]

The shoreline was only apparent at certain times and on close inspection. But anyone traveling past the lake on the main road could see the trees. The woods surrounding Thirlmere had varied greatly over the centuries, in both extent and composition, as had those of the entire Lake District. When Manchester acquired the lake they were relatively sparse—memory and legend, at least, suggested that they had been much thicker at some earlier periods—and composed primarily of deciduous trees. Some of the trees were inevitably doomed to drown under the raised waters of the reservoir. The TDA had been particularly concerned that those on higher ground be protected, not only for their own sake but as a screen concealing the disruption that would inevitably attend construction, and preservationists assumed that such protection would be included under the general obligation to respect the Cumbrian scenery that had been included in the final Manchester Corporation Waterworks Act. Not surprisingly, the Waterworks Committee did not share these priorities, and its members understood their duty of care rather differently. In the autumn of 1878, even before the purchase of the major lakeside properties had been completed, the subcommittee charged with overseeing the Thirlmere Scheme had decided not only that the woods closest to Dalehead Hall "needed thinning" but also that much of the timber throughout the manors of Wythburn and Legburthwaite should be cut. The instruction to thin "such of the woods on the Estate at Thirlmere as may be found necessary" was given the following spring.[58]

That the Waterworks Committee records routinely referred to the trees as "timber"—that is, as a product to be harvested—signaled the extent to which its evaluation of the woods diverged from that of the TDA. And even in commercial terms, assessments could vary with the occasion and the audience. An enthusiastic retrospective account emphasized the great market value of the timber, and rejoiced that the simplification of the previous complex ownership arrangements into a single large freehold meant that profit could finally be realized from a previously unexploitable resource.[59] A more nostalgic chronicler, who regretted the "winding shores . . . gracefully fringed here and there with natural woods," accepted the Waterworks Committee's characterization of the trees as "poor scrubby things, of absolutely no value as timber, and of very little as firewood," but insisted that "and yet they had a value; but no one could reckon it in shillings and pence."[60] To those who appreciated this unquantifiable value, any reduction to mere numbers, whether large or small, was distressing. The greedy "timber merchant" who "rages ruthlessly" was no less culpable than "the Manchester Corporation [who], with consummate short sight, have bared the eastern slope to the summit, and cut the woods, whose timber was of no value, smack smooth."[61] Relandscaping continued after the completion of the reservoir and pipeline, accompanied by a critical chorus that ebbed and flowed with the scale of lumbering. It reached a crescendo in the winter of 1910–1911, when the *Manchester Guardian* published dramatic photographs of clear-cutting along the western side of the lake, along with a flurry of letters to the editor.[62] One was from Gordon Wordsworth, the poet's grandson, who lamented that "what was once a sea of heather, scented by bog-myrtle, and waving with birches . . . is now a desolate and hideous flat" and that in "a relic of . . . indigenous forest" the "gnarled and lichen-covered stems . . . [of] incomparable beauty . . . have been ruthlessly felled."[63] In the *Spectator,* Thomas Thornely excoriated the Corporation in verse:

> Now false to creed and pledges given,
> Repulsing nature's kindly aid,
> She makes her artificial heaven
> In place of that which God had made.
> . . . .
> A forest—latest here that clings
> To mountainside—she tears away,
> It does not harmonize with things
> Around—suburban sages say.[64]

The enthusiasm of the Waterworks Committee for the axe did not indicate any general hostility to trees. Nor were its members concerned solely with the market price of timber; they also understood that tree cover would diminish hillside erosion and therefore help preserve the reservoir from organic contamination.[65] But they believed that some trees were more equal than others, at least for these purposes, and neither the kinds of trees at Thirlmere nor their distribution seemed optimal. Indeed, it was feared that deciduous trees, rather than helping to maintain the water's purity, would cause potential problems by messily shedding their leaves. The experience of Longdendale, where sycamores and beeches dominated the original plantation, had shown the committee how much "cost and trouble" such trees produced.[66] Therefore, even as the woods that had existed in 1878 were being cleared, Manchester was planning to replace them with the first large-scale conifer afforestation project in Cumbria. Seedlings of pine, spruce, larch, and fir were planted in regimented blocks, very different from the sparse scatter of their leafy predecessors.[67] Although boosters praised the selection of "trees that have been found most suitable to the district," preservationists viewed these alien interlopers with disdain. After the trees had had a chance to grow into dark geometrical masses, one dissatisfied observer complained that "the fir plantations about Thirlmere make the place look like a poor and small-scale imitation of the Black Forest and make one feel that it is *in* but no longer *of* the Lake Country."[68]

Open Country

When they warned against anticipated changes in the woods and waters, or when they criticized those that had already been accomplished, the opponents of the Thirlmere Scheme often appeared to be representing their own interests, or, in a somewhat grandiose metonymy, the interests of the landscape itself. This impression was heightened by their tendency to describe what they admired, and what they feared losing, primarily in visual or aesthetic terms. Since they were a relatively small group, albeit a vocal and conspicuous one, Rawnsley and his coadjutors were vulnerable to the charges of selfishness and elitism that have often bedeviled preservationists. In their campaign for parliamentary approval, advocates of the Scheme had repeatedly played the class card, weighing the reservoir that would provide Manchester's vast working population with drinking water, sanitation, and employment against the frivolous pleasures of a few wealthy connoisseurs. The defenders of Thirlmere, of course, would not have characterized themselves in those terms, and they also resisted the implication that they spoke only for themselves and their privileged kind. Especially in the wake of the railway construction that Wordsworth had so ineffectually resisted, people from many walks of life flocked to the Lake District every summer. The TDA and its supporters claimed to represent the interests of these visitors as well, and, taking a populist leaf from their antagonists' book, they emphasized the fact that many working men shared their enjoyment of the lakes and fells. As the *Times* pronounced in 1877, during the run-up to the parliamentary struggle, the undeveloped spaces of Cumberland and Westmorland were "the very lungs of the crowded cities of the North; they are to many an artizan the one opportunity of tasting purer and more elevating pleasures than those which the conditions of his daily life render possible." Pleasure in this context was hardly idle; on the contrary, it had concrete pragmatic value. Thirlmere should be spared "in the name of Utility rightly understood rather than of Beauty only, in the interest not of aesthetic *dilettanti* but of hard-worked men of business and toil-worn hands."[69] In response to the unsympathetic report of the parliamentary select committee, Farquhar argued that since "the North of England is becoming choked with mining and manufacturing works," the Lake District required preservation as one of "but few convenient places of country resort . . . where it is possible for men of small income to enjoy the fresh air and complete change of scene which is needful for them during intervals of relaxation." Rawnsley struck the same inclusive note in his later plea to establish a society that would attempt to preserve the Lake District as "one of the few recreation-grounds open to, and within easy distance for, the toilers of our northern towns."[70]

For reasons of convenience, the rural areas immediately outside large towns were the most frequent destinations for visitors of modest means. Some came for simple recreation; others' motives ranged from the nostalgia of factory workers who remembered agricultural childhoods to the more systematic curiosity of amateur naturalists. Once they arrived in the Lake District, all tended to enjoy the scenery on foot.[71] Such ramblers, as they had come to be known by the late nineteenth century, used the dense network of footpaths—traditional public rights-of-way—that crossed most private properties in the countryside. Although perfectly legal, their wanderings often irritated landowners, some of whom tried to assert control over their land by blocking the paths. Such interference, along with the erosion of accessible spaces through the continuing enclosure of common lands and manorial wastes, had prompted organized local resistance in various places—protests that often combined class consciousness with aesthetic sensibility—beginning in the early decades of the nineteenth century.[72] Thus angry residents of Whitefield and Pilkington (near Bury, Lancashire) "desire[d] not only that these footpaths [closed by the Earl of Derby] should be used, in order to shorten the journey to and from labour, but that . . . the beauty of nature may be admired."[73]

The complaints of humble pedestrians were often supplemented by more powerful voices. Threats to the open spaces of London prompted the formation of a Parliamentary Committee on the Metropolitan Commons in 1865; the passage of the Metropolitan Commons Act of 1866, which incorporated some but not all of the committee's recommendations; and the foundation of the Commons Preservation Society in 1866.[74] Although initially focused on London, the society's range of concern soon expanded, and it provided valuable, if ultimately unavailing support for the TDA's campaign. Indeed, according to founding member George Shaw Lefevre, the Commons Preservation Society could take credit for one of the consolation prizes included in the final Manchester Corporation Waterworks Act of 1879, the provision "that the access heretofore enjoyed on the part of the public and tourists to the mountains and fells surrounding Lake Thirlmere shall not be in any manner restricted or interfered with by the Corporation."[75]

Within the Lake District as a whole, footpath access proved an increasingly sensitive issue, as excursionists flocked to the scenic walks nearest the larger towns. Thus, in 1887 a landowner blocked the popular path to the summit of Latrigg, just north of Keswick, with barbed wire, sparking widespread expressions of concern.[76] As the honorary secretary of the Manchester Committee of the Keswick and District Footpath Association put it in a fund-raising circular, "Much more is at stake than the mere pathway up Latrigg . . . the ascent of every mountain height in the land will be endangered."[77] Finally two thousand protesters marched to the top, disregarding the obstacles in their way. Their action had inevitable legal consequences, the costs of which were defrayed by contributions from support groups that had formed in cities and towns throughout the nation.[78] Rawnsley helped to mediate this conflict, acting both on behalf of the Lake District Defence Society and as president of the local branch of the Footpath Preservation Society, but its resolution was not an unqualified triumph. Although access to Latrigg was guaranteed, a commentator in the *Westmorland Gazette* disparaged the outcome as "only a footpath nine feet in width, railed in on both sides—no very great victory after all."[79] In addition, the episode exposed areas of disunity among access enthusiasts, with the more radical inveighing vigorously against landlords in principle, not just when they misbehaved. It also underlined the diversity of opinion among Lake District residents. Many supported the Latrigg campaigners, but the influx of national support also inspired a resurgence of old resentment of carpetbagging aesthetes.[80]

Pedestrian access to the hills and fells surrounding Thirlmere had been guaranteed by Parliament, and, more generally, the Corporation had promised broader enjoyment of the lake as one of the benefits that would result from the implementation of the Scheme. Nevertheless, as was the case with several other provisions of the act, words turned out to be one thing and actions another. The fears satirically expressed by the author of *King Croesus; or, Harlequin Town Clerk Irwell and the Thirlmere Lake* turned out to be prescient, as did the remedy he proposed. While discussing plans to make the most efficient possible use of the lake, the engineer Hawkeye warns the town clerk that "the *sentimental party* . . . who are pleased to call themselves the Commons Preservation Society . . . will give you a great deal of trouble before they've done."[81] The town clerk replies that he will comply with the letter of the law, but not with its spirit, and for the most part, that is the policy that the Waterworks Committee followed—though occasionally legal obligations were blatantly disregarded, as when, in 1886, one of the contractors built huts and workshops that blocked "a path leading along the side of a little stream, picturesque with rock and waterfall," in the neighborhood of Grasmere.[82] Determined hill walkers were allowed to ramble (most frequently up the western route to the top of Helvellyn), but the lakeshore was posted with notices forbidding access, which could easily be misinterpreted as referring to the entire surrounding area.[83] Even after the reservoir had been completed, the lake itself remained off limits, not only to the boaters and fishers whom Heron had promised to accommodate, but to walkers and gazers as well. Decades later Thirlmere was still "unapproachable: at no point can you ramble along its shore," and the growth of conifers had made it invisible from many vantage points on the road.[84]

The main means by which the Manchester Corporation proposed to expand the potential audience for Thirlmere's charms turned out to be highway construction. New roads had formed an important component of the Scheme from its inception. The existing highway along the eastern side of the lake would be submerged by the reservoir, and the track along the western side, while affording views that Harriet Martineau thought "unquestionably finer," could not "be traversed in a carriage."[85] The plans included replacing both with new roads that would "be thrown open to the public, and . . . far more than compensate them for any supposed injury by the erection of the embankment"; the projectors felt "little doubt that . . . the circuit of the lake as it will be one of the loveliest, will become one of the favourite drives in the Lake District, whilst to the pedestrian it will offer equal attractions." James Mansergh, the designer of Birmingham's Welsh water supply, agreed with his Manchester colleagues, predicting that "the circuit of Thirlmere by means of these roads . . . will . . . become a favourite excursion, partly on account of the grand scenery, and partly on account of the 'marvels of engineering skill' to be constructed under the auspices of the Manchester Corporation."[86] Although some opponents of the Scheme might grumble that "to replace the existing bridle-path with a carriage road, after having submerged the best part of the scenery, would be an act of generosity rather difficult . . . to appreciate," many ultimately found Manchester's arguments on this single point to be persuasive. Thus in outlining a fallback position if their attempts to block the Scheme proved unsuccessful, the TDA's legal representatives planned to stipulate "that the Promoters shall be compelled to make the road round the lake," and this stipulation was ultimately included in the act.[87]

Ten years later, however, perspectives had altered. Manchester no longer seemed so eager to expose Thirlmere's most dramatic remaining vistas to the popular gaze. The Waterworks Committee had determined that Manchester would not immediately require the full potential capacity of the reservoir, and that therefore neither the water level nor the embankment needed to be as high as originally planned. In consequence, the Corporation attempted what one otherwise enthusiastic chronicler of the Scheme characterized as "the *shirking* of a most important feature . . . a carriage drive along the western shore," the promise of which had helped conciliate initial local opposition. Instead of a new elevated road, they proposed merely to modify the existing low-level track, at an estimated savings of approximately ten thousand pounds.[88]

The Corporation's habitual antagonists emerged as surprise supporters of this alteration in plans. In a June 1888 memorandum to the Waterworks Committee, the leaders of the Lake District Defence Society and the Commons Preservation Society agreed that, with the lake only to be raised twenty feet rather than the original fifty, "the proposed new high-level road . . . will no longer be necessary." They justified their unusual support of Manchester's plans in predictable terms, being "of opinion that, owing to the extensive engineering operations which would be required, the construction of a new high-level road would seriously detract from the beauty of the western shore . . . , and they believe that no adequate compensation for such injury to the scenery would be found in any added convenience for passenger traffic." They generously assumed that, like themselves, "the Corporation and people of Manchester [were] most anxious that . . . the natural features of Thirlmere should be interfered with as little as possible, and that the peculiar beauty of the lake should remain intact."[89] At a meeting of Manchester ratepayers in December 1888 Mayor William Batty stressed the Corporation's financial calculations—"the inhabitants would, if they adopted the present proposals . . . save the expenditure of a large sum of money"—but he also embraced the aesthetic rationale offered in the memorandum. The assemblage greeted both arguments with cheers of "Hear, hear" and enthusiastically approved the requested change.[90]

That was not, however, the end of the story. Since the original obligation had been enacted by Parliament, additional legislation would be necessary to modify it. As word of the proposed abandonment of the new road spread, it quickly became clear that neither the leadership of the

Lake District Defence Society nor that of the Commons Preservation Society had its finger on the local Cumbrian pulse. Even within the ranks of the Lake District Defence Society, allegiances wavered. In a letter to the editor of the *Manchester Guardian*, J. B. Baddeley, the well-known author of guides to the Lake District, announced that it was "with true regret that I find myself opposed . . . to a Society of which I have been a member from the beginning, and of which I hope to remain a member until they turn me out."[91] He particularly contrasted his views with those of H. D. Rawnsley, but Rawnsley's own position turned out to be less clear than his subscription of the original memorandum might have indicated; his name did not appear on the subsequent literature about the revised road distributed by the Lake District Defence Society. As a candidate for the Cumberland County Council, Rawnsley needed to consider the views of his fellow citizens, who greeted the news of Manchester's proposed abandonment of the promised road with howls of outrage. Or, as W. H. Hills, the secretary of the Lake District Defence Society, put it in a letter to a colleague at the Commons Preservation Society, "Great pressure was put on Mr. Rawnsley by a party of innkeepers before his election . . . and he has most unfortunately given way on important points. . . . This is nothing short of a disaster."[92] (Hills was somewhat given to unflattering private reflections on apostate coadjutors; in another letter he disparaged Baddeley as a "rather unsuccessful" schoolmaster who had abandoned his profession when his guides, designed "in close imitation" of Baedeker, had become popular.[93])

In the public debate that followed, pamphleteers on both sides invoked Rawnsley's support.[94] It was unlikely, however, that his changing views tipped the balance or even that he changed many minds; local resistance to Manchester's change of plans was more deeply grounded. Both oral and written protests emphasized perfidy, Mancunian and otherwise. One broadside featured parallel columns: on the right, a list of promises offered by numerous Manchester witnesses before the select committee; on the left, a single long quotation that emphasized the Waterworks Committee's gratified acceptance of preservationists' support for abandoning those promises. At a meeting of the Cockermouth Union Highway Authority one member spoke more plainly, accusing the Corporation of "doing their level best to sneak out of the Act of Parliament."[95] Recognizing the strength of local opposition, the Corporation ultimately withdrew its bill to amend the 1879 act. Years later, the boosterish Lake District Association, complacently (if partially) summarized the confrontation: "The Corporation evinced a disposition to evade certain of their obligations, but this Association was able to bring sufficient pressure to bear, and the road was carried out as planned, and now affords a most picturesque drive of easy gradient, and is a very obvious advantage to the district."[96] In the end, Manchester accepted this reversal gracefully. When the new road was completed in February 1894, the mayor of Manchester, attended by the mayor of Salford, the chairman of the Waterworks Committee, and more than a dozen other dignitaries, traveled to Thirlmere for the opening ceremony. Within a few years, visitors could make the circuit of Thirlmere by "public conveyance" for a fare of approximately four shillings.[97]

It had nevertheless become clear over the long period of construction that providing opportunities for casual tourists to admire Thirlmere, whether on foot, in a boat, or in a horse-drawn vehicle, was no higher a priority for the Manchester Corporation than it had been for the original defenders of the lake. And if the Waterworks Committee thus showed little inclination to accommodate visitors drawn to Lake District by its landscape, it was equally indifferent to the sensibilities of those drawn by its literary associations. Although Thirlmere was located comfortably within the territory sanctified by Wordsworth and his Romantic cohort, its immediate surroundings had boasted only one monument to their presence: the Rock of Names, a dark slab near the south end of the lake, which was said to bear their carved initials. The authenticity of the carving was occasionally questioned; one skeptic suggested that they were "the work of a rather clever but eccentric amateur stone-cutter" named John Longmuir, who had created a number of similar carvings in the neighborhood of Windermere and Ambleside. But by the time the Thirlmere Scheme took shape the Rock of Names had become a station of literary pilgrimage, a "memorial of the tryst of the poets" that was "carefully guarded by the moss and the lichen, unknown save to readers of the bard."[98] Indeed, the poet's son William Wordsworth had mentioned it to the select committee as a compelling reason not to flood Thirlmere.[99]

His testimony had negligible impact, however, and the rock was doubly doomed by the reservoir construction. Not only did the plans call for it to be submerged beneath the expanded lake, but one of the contractors proposed to blow it up and use the fragments to build the dam. The Waterworks Committee made no effort to avert this fate until several antiquarians offered to buy it; ultimately Harwood (who suggested his own skepticism when he described the initials as "said to be carved at a time when a pic-nic was held at Thirlmere, which the poets and their friends attended") agreed to give it to the Wordsworth Institute at Cockermouth,

The Rock of Names reaggregated.

the poet's birthplace. The rock proved impossible to move, however, so in the end only some of its face was saved, and the contractor had his way with the rest. Even in this disaggregated condition, much of its charisma remained—in 1894 a visitor recorded with only gentle irony his companion's desire to view "the sacred 'Rock of Names,' or all that remains of it."[100] What remained was a cairn, erected near the new high road, described by the ubiquitous Rawnsley as "carefully built" and containing "only the fragments of certain letters, which are all that we [were] able to save from the cruel blasting powder of the contractor." One traveler suggested that it had been symbolically replaced by "that modern 'Rock of Names,' the hideous granite slab . . . [erected by the embankment at the north end of the lake] on which the names of sundry members of the Manchester Waterworks Committee are enduringly engraved."[101]

Abundant Waters

Construction proceeded steadily despite continual surveillance and intermittent protests, and by the summer of 1894 the end was in sight. All that remained was for the reservoir water to begin flowing through the pipeline. But at this point the Waterworks Committee hesitated, not in response to any of the objections raised by critics over the preceding fifteen years, but in response to a more conventional municipal consideration. After the works were officially completed, the Corporation would have to begin paying local taxes, characterized by Harwood as "heavy," at the reservoir site and along the whole route of the aqueduct. Naturally it was anxious to delay such payments as long as possible, and so its inclination was to delay the opening while the need for additional water was not acute. Manchester's hand was forced, however, by public concerns about the quality of the extant water supply. During the dry summer of 1893, the Waterworks Committee had had to supplement Longdendale water with noticeably inferior water stored in other city reservoirs.[102] The Corporation feared that if this situation recurred in conjunction with an outbreak of contagious disease, the water supply would be blamed, whether justly or not—by this time the general connection between epidemics and contaminated water was well established. Therefore, in August it decided to proceed with arrangements for an opening ceremony at the reservoir. The precise date was not fixed until the following month, in the hope that the Prince of Wales would accept an invitation to attend. When he politely declined "on account of his numerous engagements," the Waterworks Committee decided that, in addition to the formal opening of the reservoir and pipeline at Thirlmere on Friday, October 12, its urban constituency should be given a chance to enjoy a "public demonstration" the following day.[103]

Despite the shotgun timetable, both ceremonies were magnificent. (Manchester had had recent practice with staging magnificent ceremonies. Grand as they were, the Thirlmere reservoir and pipeline were not the most important public works inaugurated by the city in 1894. At the beginning of that year Queen Victoria had traveled to Manchester to open the thirty-five-and-a-half-mile-long Ship Canal, which linked Manchester to the River Mersey near Liverpool, and thus to the global oceans.[104]) Since the Thirlmere observance marked the official opening, it was the more elaborate. Nearly one hundred members of the Corporation and other dignitaries traveled by special train to the rail terminus at Windermere, and then in a procession of over thirty carriages to the reservoir.[105] The *Manchester Guardian* reported that they received a cordial welcome as they drove, suggesting that locals had finally become reconciled to the Scheme: "At Ambleside large numbers of people congregated to see the procession pass," and at Grasmere the inhabitants showed their

ALDERMAN SIR JOHN JAMES HARWOOD.
Ex-Mayor of Manchester.
(From a Photograph by M. Gutenberg.)

"goodwill . . . in a display of bunting." "A large crowd of people resident in the district" also came to witness the formalities.[106] As chairman of the Waterworks Committee, Harwood was the star of the show at the crenellated Gothic Revival straining well. Before turning the mechanism that allowed the reservoir water to flow into the aqueduct, he gave a long, complacent speech in which he recalled the initial resistance to the Scheme. "Is there anything hideous in the handsome embankment we have formed?" he asked. "Is there one of you who thinks that jaded men and women, who seek restoration to health and strength in the quiet contemplation of the unspeakable beauties of Nature, will be deterred . . . because we have applied this watershed to the purposes of civilisation?"[107] He immediately provided his own answer—"No; a thousand times no"—but his questions were not entirely rhetorical.

The ceremony was designed to symbolize the human links between Manchester and the Lake District, as well as the physical connection by means of the pipeline. The invited guests therefore included members of the former TDA, and one prominent activist even shared the platform. Wearing his clerical hat, Rawnsley opened the proceedings with a prayer that "this river of God flow through the far-off city to cleanse and purify; to help and heal." At the elaborate ritual meal that followed the ceremony he offered an extended toast to "the Waterworks Committee of the Manchester Corporation," comparing the fruits of their hydraulic efforts favorably to those of ancient Rome and, somewhat more opaquely, comparing Harwood's extraction of water from Thirlmere to King Arthur's extraction of Excalibur from a stone.[108] Prolific in verse as well as in prose, he also composed four sonnets for the occasion, which suggested a more complex and nuanced response. Three were extended and appreciative apostrophes: "To the Promoters and Builders of the Thirlmere Waterworks" ("Rome sets the seal of her imperial power / To this your bold endeavour . . ."); "To Sir John Harwood" ("For you had gauged the future, felt the stress / Of that great city's toil and thirst and strife, / . . . And claimed from Cumbrian hills their gift of life"); and "To the Workmen—Those who live, and those who died—who brought the Thirlmere Water to Manchester." The fourth, entitled "Thirlmere: Loss and Gain," weighed the altered shoreline, the diminished grandeur of Raven's Crag, the banished inhabitants, and the shattered Rock of Names against the clearer view of Helvellyn from the western shore and the more abstract possibility that "Peace secure can walk the mountain ground / Love consecrates to serve the common weal." Although Rawnsley claimed that these were "compen-

COMMEMORATIVE SONNETS.

BY REV. H. D. RAWNSLEY, M.A.

To the Chairman
and the Members of the Waterworks Committee
of the Manchester Corporation, the Engineers,
Builders, and Workmen, on the occasion of the
Opening of the Thirlmere Waterworks.
Friday, 12th October, 1894.

To Sir John Harwood.

OUR generations pass, our names decay,
The spirit lives—the courage and the skill
That chains the torrent, that can pierce the hill
And lead sweet water-floods from far away;
And tho' this vast laborious essay
May fade in new adventure, men shall still
Honour the heart and praise the stubborn will
That brings strong purpose to its goal to-day.

For you had gauged the future, felt the stress
Of that great city's toil and thirst and strife,
And when men's tongues were clamorous and loud,
You held high commune with the silent cloud,
You trusted wide Atlantic waves to bless,
And claimed from Cumbrian hills their gift of life.

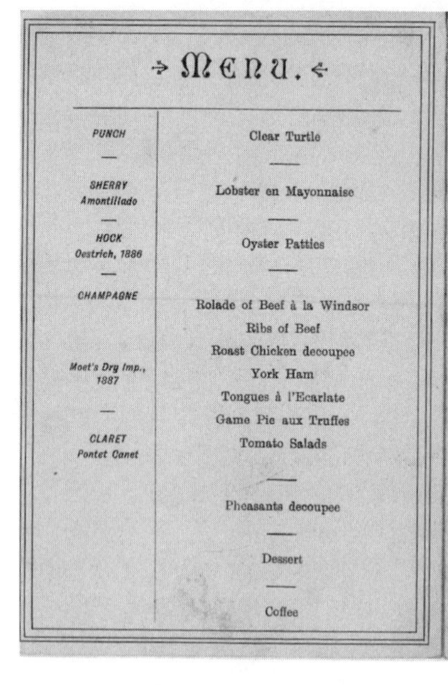

→ Menu. ←

PUNCH	Clear Turtle
—	—
SHERRY *Amontillado*	Lobster en Mayonnaise
—	—
HOCK *Oestrich, 1886*	Oyster Patties
—	—
CHAMPAGNE	Rolade of Beef à la Windsor
	Ribs of Beef
	Roast Chicken decoupee
Moet's Dry Imp., 1887	York Ham
	Tongues à l'Ecarlate
—	Game Pie aux Truffes
CLARET *Pontet Canet*	Tomato Salads
	—
	Pheasants decoupee
	—
	Dessert
	—
	Coffee

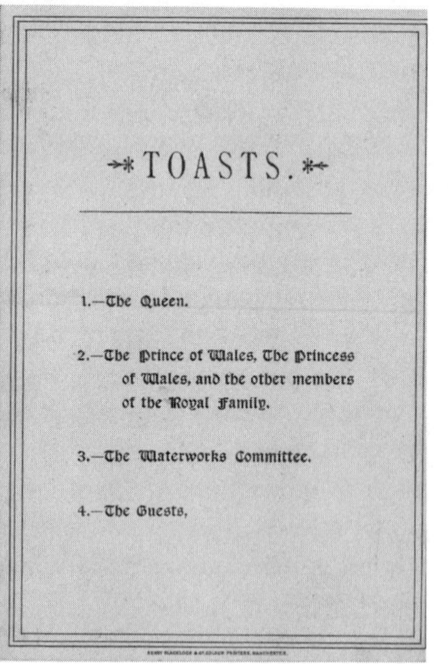

→* TOASTS. *←

1.—The Queen.

2.—The Prince of Wales, The Princess of Wales, and the other members of the Royal Family.

3.—The Waterworks Committee.

4.—The Guests.

TEMPORARY FOUNTAIN IN ALBERT SQUARE—DISPLAYING FIRST WATER FROM THIRLMERE.
R. Banks] [*Manchester.*

sations for the wound" inflicted by the Thirlmere Scheme, he also, and perhaps more persuasively, described the injury as "too deep for any time to heal."[109]

At Manchester the next day there was no such ambivalence, and the solemnity that had characterized the ceremony at the reservoir was replaced by a festive holiday atmosphere. Even the weather proved "exceptionally favorable," in contrast to the "thick and misty atmosphere" that had prevailed in the Lake District.[110] A dense, good-humored crowd filled the large Albert Square in front of the town hall, where a fountain had been erected for the occasion; other viewers jammed nearby roofs and windows or, more precariously, straddled the projecting stonework that adorned the municipal headquarters. While waiting for the program to begin, the crowd was entertained by the town hall bells and the police band. Most of the dignitaries who had attended the ceremony in Cumberland reappeared, even though their train had not brought them back to Manchester until late the previous evening. They heard speeches that were equally celebratory and more triumphalistic in tone. The concerns of the early opponents of the Scheme were mentioned only to be dismissed. As the mayor rather smugly

OPENING OF THE THIRLMERE WATER WORKS.

CITY OF MANCHESTER.

The Lord Mayor and Corporation
request the honour of the Company of
Mr C. W. Sutton
on Saturday, the 13th October, 1894.
3.15 p.m. Opening Ceremony.
4.30 p.m. Dinner in the Large Hall.

Morning Dress
Town Hall, Manchester.

Carriages at 8 p.m.
An early reply will oblige.

This Card to be presented at the Princess Street entrance not later than 2.45 p.m.

put it, "Of course the inhabitants of that district did not desire to see their country disfigured, but they forgot, what . . . they ought to have taken into consideration, the object that Manchester had in view. Sentimentalism . . . ought to have given way in the face of the necessity of conferring upon a large and crowded population the inestimable boon of a good supply of water."[111] The completed reservoir and pipeline were extolled as "victories of peace," as well as of hygiene and progress. Again, Harwood set the machinery in motion, this time "in the character of a genie of the Arabian Nights," with a specially crafted golden key that unlocked the fountain so that "the first jet of Thirlmere water, fresh from its long journey . . . from its mountain cradle, shot gaily and resolutely aloft."[112] The speeches were interrupted by repeated cries of approval, and the water was still more enthusiastically greeted; some onlookers had even brought cups so that they could catch some of the virgin drops and drink them. While the general populace continued to celebrate in the streets, several hundred invited guests retired to the town hall for an even larger and longer ceremonial repast than the one that many of them had eaten the day before, its culinary preeminence signaled not only by two additional courses but by the fact that menu was printed in French rather than English.[113]

∵ MENU. ∵

Vins.	
CHABLIS	Huitres au Citron
—	
PUNCH	POTAGE Tortue Claire
—	
HOCK	POISSON Filet de Turbot à la Homard
—	ENTRÉES Vol au Vent à la Financière Filet de Venison à la Kinnaird
CHAMPAGNE Bollinger's 1887	RELEVÉS Selles de Mouton à l'Anglaise Dindon Pique à la Toulouse et Langues de Bœuf
—	
AMONTILLADO SHERRY	Jambon de York au Champagne
—	GIBIER Faisans Coq de Bruyère
CLARET Smith Haut Lafite	ENTREMETS SUCRES Gelée au Madère Turban de Abricots à la Crême Gelée au Noyau Rouge Petites Gateaux à la Royale
—	
PORT	
—	
LIQUERS COFFEE CIGARS	Pouding Glacé *DESSERT.*

∵ TOASTS. ∵

1.—The Queen.

2.—The Prince and Princess of Wales, and the other Members of the Royal Family.

3.—Sir John J. Harwood (Chairman of the Waterworks Committee).

4.—Mr. G. H. Hill, C.E.

5.—The Guests.

The Eye of the Beholder

The impressiveness of the completed works was never in question. In both popular and specialist accounts of the engineers' technical achievement, the "transformation, or at any rate, modification of Nature's work," was routinely evoked in terms of its awe-inspiring dimensions, whether length of pipeline, or volume of water, or area of lake, or breadth of dam (or, in a slightly different register, cost of project).[114] Appreciation transcended national boundaries. Thus a weekly French magazine of popular science provided a litany of statistics in its laudatory report on the Scheme. It also reviewed the history of the project from a perspective wholly consistent with that of the Corporation, beginning with a characterization of the TDA's parliamentary opposition as "political intrigues" and "machina-

La grande digue du lac Thirlmère. — La chaussée carrosable.

The completed embankment.

Looking across Thirlmere at Helvellyn

tions" which ultimately "yielded to reason."[115] Australian admirers offered more lasting homage by rechristening the Picton Lakes (earlier known as Couridjah Lagoons) in New South Wales as the Thirlmere Lakes; the name of a nearby settlement was changed from Redbank to Thirlmere.[116]

Despite Harwood's complacent answer to the questions he posed at the lakeside ceremony, there was little consensus about other consequences of the lake's transformation. Almost every early prediction turned out to be vindicated from one perspective or another. Debate about the finished reservoir was, therefore, nearly as heated, and opinions at least as various, as they had been when the Scheme was first mooted. One admirer of the original lake conceded that the revised version was "a magnificent, possibly even a very beautiful reservoir, but not Thirlmere"; on the contrary it was "death"—"the death of plethora, not of exhaustion—but death all the same." Another judiciously described himself as "not one of those who complain that it has very little beauty left," before asserting that "it has lost certain charms since the Manchester Corporation set to work on it."[117] An elaborately illustrated guide to the Lake District acknowledged that from some perspectives, especially those that obscured "the sham castle that Manchester erected . . . a pretentious battlemented horror of red sandstone" (that is, the straining well), Thirlmere appeared "exceedingly pretty . . . and you need not remember that the mere is semi-artificial." A cycling enthusiast considered the flat new road "a benefit which goes far to make up for the water towers and other slight disfigurements." The novelist Mrs. Humphrey Ward denounced "the wrack and ruin which Manchester has made of the once lovely shore of Thirlmere."[118] The continuing force of arguments that could be disparaged as aesthetic or sentimental was shown by the frequency with which they were addressed, even by commentators inclined to praise the technological prowess of the engineers or the foresight of the municipal planners. For example, the enthusiastic French reporter acknowledged that although "the lake had lost its gracious, smiling aspect, it had gained a severe majesty" in compensation.[119] An official account explained that the ultimate fifty-foot rise in the level of the lake would constitute "an artificial restoration" of its former contours, and that, as a result, "in the new Manchester Waterworks, Nature has been supplemented by art."[120]

When they were assured of a sympathetic audience, however, eminent Mancunians felt less need to pull their punches. In 1914, for example, a group of city council members traveled to Thirlmere to admire the progress of afforestation and to consider raising the water to its maximum height. After inspecting the reservoir, they enjoyed a convivial dinner at a Grasmere hotel, at which the chairman of the Waterworks Committee declared that "the only fault" of the Scheme "was that it was not big enough."[121] But by this time, invidious comparisons were becoming less frequent. The original defenders of Thirlmere had striven to protect the lake as they knew it in the mid-Victorian period, not the lake as it had been a century earlier. With each succeeding decade fewer people remembered the undammed lake. By 1910 a brochure published by the London and Northwestern Railway to promote its services to Windermere, Kendal, and Keswick could state flatly that "Thirlmere . . . has not been, as it was suggested it would be, turned into an abomination by the Manchester Corporation"; on the contrary the lake and its surrounding hills provided "a continuous feast to the eye."[122] A contemporary motorist agreed that "the Manchester ownership—prosaic as it may be—has not detracted from the beauty of Thirlmere, and many [unspecified] nuisances that once encumbered its shores have been abated."[123] Ten years later it could seem completely naturalized, at least to the untutored gaze. An enthusiastic American tourist found that "a four-horse touring coach, with red-coated coachman and guard . . . added a touch of gaiety . . . to the beautiful but otherwise lonely scene" and was surprised to learn that "this little lake, absolutely unspoiled in itself and its surroundings, is the water supply for the city of Manchester."[124]

With Sir J. J. Harwood's compliments

5

The Harvest of Thirlmere

From the Manchester perspective, the ceremonial reception of the first Thirlmere water formalized a relationship that had been developing for decades. Although the official connection was based on the purchase of land (and water), it involved much more than the exchange of deeds for money. When the Manchester Corporation acquired the two large properties at the heart of its Thirlmere estate, it replaced the head of the Vane family as lord of the manor of Wythburn and the head of the Leathes family as lord of the manor of Legburthwaite. However odd—even oxymoronic—it may have seemed for the urban symbol of modern manufacturing and commerce to fill this medieval role, the Corporation undertook its new responsibility with earnest enthusiasm, albeit at one remove, insofar as the lord of the manor had to be a flesh-and-blood person. Thus representatives were delegated who could, if required, preside over a manorial court.

The lords of Manchester's rural manors were chosen from among the prominent members of the city council. Fittingly, the first to occupy this position was John Grave, the native Cumbrian who had originally persuaded the council to choose the Thirlmere option; an 1879 document appointing a steward of the manor of Legburthwaite described Grave as "a Trustee for the said Mayor Aldermen and Citizens [of Manchester]."[1] He enjoyed his lordship, especially as it offered him a position of continuing municipal influence, when, soon after the parliamentary approval of the Thirlmere Scheme, he had to relinquish the chairmanship of the Waterworks Committee. His ouster had nothing to do with the committee's aggressive acquisition of Lake District property, nor did it reflect any corporate second thoughts about the reservoir site. Rather, an inquiry into slack accounting practices at the Waterworks Department had revealed that Grave, demonstrating the trademark enterprise and opportunism that had guaranteed his success in business, had repeatedly used its funds as a convenient source of interest-free loans. He turned out to have similar difficulty distinguishing between public and private interest with regard to his manorial lordships, considering them his personal prerogative and resisting vigorously when the council decided that it was time to replace him.[2] His intransigence reflected ambiguities in the arrangement, as well as personal predilection. The formal language used to record manorial activities may have encouraged Grave's inclination to conflate self and city. Not every document specified the subservience of the lord to the municipality, and some explicitly gave the opposite impression. For example, when Harwood assumed these positions in 1891, the Court Book of the Manor of Wythburn stated that the "General Fines upon the several . . . Customary Tenants" were due to him personally, with no mention of the city council, and when, ten years later, he appointed a manorial steward, he did so in the name of "Sir John James Harwood, Knight my heirs and assigns," rather than in the name of the Manchester Corporation.[3]

Despite such potentially seductive rhetoric, Harwood labored under no misapprehension about the nature of his claim on the office and, perhaps for this reason, no one tried to deprive him of it; it passed peacefully to Bosdin Leech, another longtime city councillor and former mayor, at his death in 1908. Harwood felt a special closeness to the reservoir and the surrounding area. By 1894 he was, as he noted in the preface to his detailed account of the completed Thirlmere Scheme, "the only member now living of the Sub-Committee who were entrusted with the duty of making the first purchases of land at the Lake."[4] The politicians, engineers, and managers who had seen the project through to the end shared a similarly grounded sense of proprietorship. But as the enthusiastic celebrations of

October 1894 demonstrated, ordinary Mancunians also felt a bond with the distant Lake District—a bond that reflected affection as well as utility and economics. (And they were not alone in this sense of easy familiarity. Along with the waterworks, newspapers repeatedly documented the progress of a racehorse named Thirlmere, a steamship named *Thirlmere,* and *My Lord in Livery,* a farce featuring Lord Thirlmere in the title role.[5]) The municipal bond was constantly reinforced by official reminders of the source of the water that made daily life safer and more pleasant. The reservoir lent its name to avenues, roads, streets, drives, and mews throughout greater Manchester. The Waterworks Department distributed leaflets that described the Scheme along with those that specified the rates for using municipal water, offering the impressive statistics arrayed in the former as implicit justification of the elaborately scaled charges itemized in the latter (for example, the cost of watering a garden depended on its area, while the cost of washing windows depended on the value of the premises).[6] School presentations and exhibits designed for the general public rehearsed the story of the creation of the reservoir as "one of the romances of municipal enterprise."[7] Every citizen was thus offered a share of paternalistic pride in this heroic achievement.

Thirlmere, along with its surrounding watershed, had effectively become a colony of Manchester, and, transposing the figurative to the literal, some Mancunians began to colonize it. Dalehead Hall, the former seat of the Leathes family, provided summer recreation for civic worthies, as well as the local headquarters for reservoir operations. The city's ruling elite could thus enjoy the water of Thirlmere at both ends of the pipeline—drinking it at home and admiring it from the comfort of the lakeside manor house. As H. D. Rawnsley put it, striking his signature note of nostalgic resignation, the "beautiful old house with its great dark oaken stair and its wondrous kitchen fire-place is in the hands now of the Manchester Corporation, and is still hospitable as of old. But the days are gone when the quiet afternoons upon the garden lawn . . . were full of the fun of Robert Southey and the serious talk of the Rydal poet."[8] (His nostalgia was, of course, vicarious. Rawnsley had arrived in the Lake District much too late to experience these convivial glory days.) The Corporation developed a similar (and similarly nostalgic) fondness for Dalehead Hall, retaining it as a lakeside retreat for the Lord Mayor and as the occasional site of important functions until late in the twentieth century, even after it had ceded possession of the reservoir itself.[9]

Some Corporation members went further in concretizing their rela-

Dalehead Hall today (now a hotel).

tionship to the source of their water supply. Land in the immediate vicinity of the reservoir was part of the Manchester reservation and therefore not available for private purchase, but it was possible to buy and build close to the route of the pipeline. Thus, at about the time that the reservoir was completed, Edward Holt, a member of the Waterworks Committee and a future mayor, devoted some of his beer-based fortune to constructing Blackwell, an elegant arts and crafts house with a magnificent view of Windermere. Although he originally intended it as a holiday residence for his family, Holt became sufficiently engaged in Cumbrian affairs to serve as a local justice of the peace.[10] Those who merely desired accommodation, rather than possession, could live closer to the lake. As the population of Thirlmere's immediate surroundings dwindled, Waterworks employees came to constitute an increasingly large proportion of it. The 1851 census had counted 203 residents of Wythburn alone. In 1901, a local directory listed 17 residents (heads of households) in Wythburn and Legburthwaite combined, of whom only seven were farmers. The remainder included the straining well engineer, the waterworks inspector, the road foreman, and the Corporation's housekeeper at Dalehead Hall.[11]

Constant Comment

The bond connecting the reservoir to Manchester was also felt strongly in the Lake District, but the relationship was asymmetric. When a hill-walking guide described Thirlmere as the "bath of Mancunians," it emphasized the oppression latent in domestic closeness.[12] Ownership of the enormous Thirlmere watershed gave Manchester the local influence and responsibility that fell to any substantial rural landlord, and both influence and responsibility were intensified by the continuing impact of the Waterworks Committee's activities. Many of the controversies that had emerged during construction resurfaced repeatedly, especially since the completion of the reservoir did not mean the end of work on the Scheme. In 1894 the water level of the lake had been raised only twenty feet, and four of the five projected pipelines remained to be laid.[13] Aqueduct construction began again in 1902 and proceeded intermittently until 1925, when the fourth pipeline was completed (it was larger than originally planned, so the fifth was not built). In 1918 the reservoir embankment was raised three and a half feet, several years in advance of the requisite parliamentary approval.[14] Even the normal operations of the waterworks could resurrect old arguments. For example, Manchester's management of the water level within the drainage area (not just in the reservoir itself) provoked a range of grievances that varied with the local rainfall. Thus in 1924 Cumbrians complained, on their own behalf and on that of the local fish, that the large amount of water extracted by the waterworks had depleted the Thirlmere watershed; the next year, conversely, the waterworks were blamed for local flooding.[15] Critiques of the lake's unsightly exposed shoreline reemerged whenever the water level fell during dry summers.

Nor did the routinely efficient and innocuous operation of the waterworks allay the profounder fears that had troubled many locals before the reservoir opened. Early opponents of the Thirlmere Scheme had suggested that either the dam or the pipeline might fail, concerns derived both from the spectacular failure of less ambitious water projects and, it was snidely suggested, from the engineer Bateman's previous technical lapses in the construction of the Longdendale reservoirs. Thus, in preparation for the select committee hearings, the Thirlmere Defence Association's barrister was advised "to cross examine Mr Bateman as to the faults and other subterranean dislocations and disturbances which may affect the stability and tightness of this Reservoir."[16] The suggestion of incompetence evidently rankled, since Bateman continued to defend the geological integrity of the Scheme even after it was approved and under construction, asserting, for example, that "the aqueduct . . . will be cut out of the solid ground, so as to be free from all risk of leakage or bursting."[17] Nervous Cumbrians were not alone in these fears, and remote but widely publicized disasters like the Johnstown (Pennsylvania) flood of 1889 only heightened them. Despite authoritative assurances that "any competent engineer can avoid the mistakes" that had led to the more spectacular contemporary dam failures, both producers and consumers of waterworks continued to worry.[18] From a technical perspective they had real cause for concern; in the pioneering era of heroic dam construction, expertise did not always match ambition. Responsible Victorian engineers compensated for limitations in their knowledge of their materials (concrete and masonry) and in their ability to calculate stresses by overbuilding their dams.[19] Bateman and Hill were responsible engineers, and no failure occurred at the Thirlmere embankment.

This strategy worked less well for pipelines, however, and in fact the Thirlmere aqueduct suffered several dramatic breaks.[20] In 1899, for example, the pipe (at that time the only one) burst where it traversed a Lancashire farm. According to the *Times,* "The accident . . . was marked by a loud report, earth and rock being hurled high into the air, followed by huge column of water, which . . . swept with devastating force over the adjoining fields." In 1935 one of the pipes similarly gave way on a hilltop near Kendal, creating a crater twenty-three feet in diameter and fifteen feet deep, as "hundreds of tons of soil and stones were swept down the hillside by the torrent," burying nearby fields, hedges, and roads.[21] Although such breaks were conspicuous, workers quickly repaired the damage. (That is, they quickly restored the flow of water to Manchester; the violently disrupted local landscapes took longer to recover.)

A break in the dam would, of course, have been catastrophic rather than inconvenient. Even its established safety record could not calm all fears, since past was not necessarily prologue with regard to structural integrity. In any case, intrinsic weakness did not provide the only possible hazard. An enormous, artificially contained body of water offered a tempting target for external assault and therefore became a magnet for more generalized anxieties. The cigar box packed with explosives, found in the course of a routine pipeline inspection in 1921 and attributed to the Irish nationalists of Sinn Fein, caused no physical harm but nevertheless confirmed the waterworks' vulnerability to terrorism.[22] In 1937, as war in Europe loomed again, the Thirlmere Parish Council discussed the possibility of an aerial attack. The reassurance given by the chief engineer—that Manchester had supplementary storage capacity near the city, which would meet its needs in such an emergency—revealingly missed the point of its concern. Altruistic Cumbrians may also have been pleased to learn that the Air Raids Precautions Committee was looking into the situation of towns along the way, which would lose their water if Thirlmere were bombed, and which had no such handy reserves. But the urban authorities offered only the coldest comfort to the more self-interested residents of a rural area that had been transformed into a military target by the dam: "The surrounding hill-tops are constantly hidden in mists . . . there are dangerous air currents . . . the dam is an extremely small target . . . There are trees giving cover . . . a dam such as this is no flimsy thing which can easily be destroyed."[23]

The Forest and the Trees

Whatever their potential efficacy (fortunately never tested) as a defense against the Luftwaffe, the Thirlmere woods inspired continuing debate. Sometimes contention focused on the fate of individual trees. When a grove of Scotch firs near Wythburn was cut down, a self-professed lover of Lake District scenery (a mathematician at the University of Manchester) mourned their lost beauty, while a member of the Waterworks Committee riposted that "half of them were dead and the remainder dying," adding insult to injury by listing the national awards bestowed on the Thirlmere conifer plantations.[24] Where conifers were concerned, however, aesthetic interests more frequently favored demolition. Even some forestry enthusiasts felt that Corporation policy had weighted landscape too little and lumber too much; thus at the annual dinner of the English Arboricultural Society in 1925, Viscount Ullswater complained that the lake was "hidden behind . . . countless evergreens. . . . For the public, it might as well be non-existent."[25] More predictably, in 1926 a deputation from the Lake District Defence Society met with the chairman of the Waterworks Committee and the head Manchester forester to protest "the rapidly increasing concealment of Thirlmere by the growth of the conifers between the lake and the road." Eager to avoid the charge of elitism, the protesters emphasized that "they had received many appeals from the general public as well as from their own members." Their specific request was modest—not wholesale clearance, but the creation of a few scenic gaps in the trees—but they received only a promise that the committee would "give sympathetic consideration to the request."[26] They were also assured that new plantation, which had been carried on at a rate of about 150,000 seedlings per year, was at an end.

The seedlings already planted, however, continued to grow taller and

Tree-felling at Thirlmere.

thicker, darkening the landscape and obscuring the lake. The unsatisfactory result of the Waterworks Committee's "sympathetic consideration" was suggested by the reiteration of similar complaints across the decades. In 1935 H. H. Symonds, the headmaster of an independent boys' school in Liverpool as well as a vociferous countryside enthusiast, disparaged "those monotonous and level tones of green" that dominated the Thirlmere slopes. A decade later he was no better pleased by the committee's attempt to "use the space 'productively,' but not to obstruct the view," calling the plan to cut "larger gaps through the trees" and replant "these gaps with Christmas trees, which will be cut in December as they reach a height of 3 feet or so . . . a queer bag of tricks." Not surprisingly, the Corporation evaluated this commercial venture differently, explicitly praising it in a publication produced to celebrate the centenary of the Longdendale waterworks.[27] In 1959 the Lake District Planning Board respectfully suggested that although "Manchester could hardly be expected to encourage visitors to halt and picnic alongside its drinking water . . . [it] need not fear that they would cast the evil eye upon it," and requested that the "thick belt of conifers" that hid the lake from the road be thinned. In this case, consideration proved effective as well as sympathetic; several months later the planning board happily reported that "the corporation was determined to see that the appearance of the reservoir and surrounding woods was made as beautiful as possible" and that "visitors would now gain fresh views of the lake."[28]

For most critics of Manchester's aggressive arborism, resistance to the spread of evergreens at Thirlmere was part of a larger struggle. Cumbrian landowners had been planting and replanting their woodlands for centuries, but the nature and pace of afforestation changed in the early twentieth century.[29] The large-scale introduction of conifers around the reservoir was a very heavy straw in a strong wind, anticipating the more widespread activities of the official Forestry Commission, which was established by Parliament in 1919 with the twin goals of "providing at home a substantial reserve of growing timber and of promoting employment in rural districts."[30] In the Lake District, the commission commenced its activities in the hills around Keswick.[31] Resistance to the spread of dark geometrical masses like those that had transformed Thirlmere's slopes coalesced in the 1930s, to protest the threatened afforestation at Eskdale and Dunnerdale (Duddondale) in the western part of the region.[32] In response, the Forestry Commission "recognised that large-scale afforestation and the preservation of areas of typical natural beauty are both necessary on national grounds, and that the land available for both purposes is limited"; it expressed the hope that "both national interests . . . can be best secured not by rivalry or obstruction, but, where possible, by cooperation or at least by a just correlation wherever otherwise the two would conflict."[33] Although afforestation proceeded in the contested areas, the commission promised to exclude the heart of the Lake District from its future purview, but neither this promise nor the measured ambiguity of the commission's commitment to compromise offered reassurance to Symonds, whose published critique found an enthusiastic audience. Reviewers particularly praised his characterization of the commission as "an entirely irresponsible body," whose members were "deaf to every plea, regardless of the most influential public opinion, contemptuous of every argument."[34]

Symonds was not opposed to afforestation in principle or in every place, but only to the form it would inevitably take in the Lake District. Since forestry was conceived primarily as a commercial agricultural enterprise, the tree species to be cultivated would be chosen according to very practical criteria. Thus, he foresaw that "any large-scale afforestation of the higher levels of the valley sides and fells of the Lake District must, by reason of soil and climate, be of necessity with conifers."[35] Both past and future experience proved him correct. Although every protest mentioned the superior desirability of deciduous trees, on grounds of aesthetics and of heritage, economic considerations inevitably inclined foresters toward evergreens. And as time went on many eyes became accustomed to the altered appearance of the landscape. Afforestation advocates began to cite the trees themselves as scenic attractions, so that Thirlmere became "noteworthy for extensive plantations of Douglas fir, pine, spruce, and larch," and for "the fine impression of the grandeur of its tall conifers" as seen from the road.[36]

Less partial opinion remained divided—"one man likes it, another dislikes it," as a post–World War II account put it—and resistance to what Alfred Wainwright, the celebrated chronicler of Lake District walks, called "the green shroud" never completely disappeared.[37] By the late twentieth century the arboreal tide was beginning to turn. Susan Johnson, the daughter of H. H. Symonds, and thus linked to earlier protest by ties of blood as well as of affinity, successfully sued the North West Water Authority (the successors to the Manchester Corporation after the nationalization of water supply in 1973) on the grounds that extensive conifer afforestation violated the requirement in the 1879 Act that "all reasonable regard shall be had to the preservation . . . of the beauty of the scenery." The consequence was a slow increase in the planting of deciduous trees and native shrubs.[38] Even the Forestry Commission ultimately experienced a change of heart. In 2005 it announced plans to remove the large conifer plantations that had inspired the controversy of the 1930s, and to replace them with "a native birch and oak woodland" in the hopes of increasing biodiversity and encouraging the return of such species as "dormice, wood ants, redstarts and pied flycatchers."[39]

The Parts and the Whole

Despite these limited reversals, many of the ecological changes caused by afforestation had become well entrenched. By the middle of the twentieth century, some plantation conifers, including the lodgepole pine, the Sitka spruce, and the European larch (introduced in the eighteenth century but with a comparatively limited distribution) ranked among the most common Cumbrian trees.[40] They joined a varied assemblage of nonindigenous plants that had found hospitable habitat in the Lake District. Nineteenth-century botanists had acknowledged the number and variety of these interlopers when they cataloged the local flora, supplementing standard binomial taxonomy, which was based on physical similarity and common descent, with a set of categories that highlighted geography and heritage. These additional categories, termed "classes of citizenship," identified plants as *natives,* which had established themselves "quite independently of man's intervention"; *colonists,* "well-established weed[s]" that grew in arable land; *denizens,* which presented a deceptively wild appearance "but may perhaps have been originally introduced through human agency"; and *aliens,* which "without doubt [had] strayed from cultivation."[41] A turn-of-the-century survey pointedly suggested that most Cumbrian aliens throve in the vicinity of harbors.[42]

At issue were both indigenousness and wildness, qualities that overlapped but were not identical. As with many classification systems, differentiation implied a hierarchy that could be applied to classifiers as well as to their subjects: it was "the fault of botanists of small experience, or sometimes of small conscientiousness," to confuse aliens with (implicitly superior) natives, by "placing them on a level with the really wild plants."[43] The distribution of plants was characterized in terms that were merely geographical and therefore somewhat less evocative. But they also valorized neighborhood provenance, descending from the "Local Type" through the "Intermediate Type.—Species that have their headquarters in the north of England" to other "Types" more remotely or more generally headquartered, including the Scottish, the Germanic, and the Atlantic.[44]

In the course of the twentieth century, the "distinct botanical individuality" with which a late-nineteenth-century botanist had credited the Lake District diminished markedly.[45] The number of native species declined, even though the discovery of previously unnoticed species compensated for some of the disappearances, and the number of aliens increased by over 50 percent.[46] The flora of the Thirlmere watershed showed a similar shift in proportion. This local decline in native species could not, however, be completely attributed to invaders, whether intentional human introductions or enterprising independent migrants. Raising the water level of the reservoir drowned the established lakeside plants, and the repeated exposure and submergence of the new shore's muddy shingle discouraged their reestablishment higher up.[47] The reengineering of the lake also influenced the distribution of vegetation in its neighborhood, as well as on its shores. Its expanded surface area produced subtle alterations in local microclimates, and the mandated cessation of traditional economic activities—especially the disappearance of most of the sheep—had a still more obvious impact.[48] As the environment changed, some plants that had been well adapted to previous conditions could no longer thrive.

The reservoir also influenced faunal variety, although not necessarily for the same reasons. At the end of the nineteenth century foxes, badgers, otters, pine martens, stoats, weasels, several species of deer, and a varied mix of rodents, insectivores, and bats all lived in the Lake District. Because of the region's remoteness from centers of human population and its unsuitability for intensive development, remnant populations of some species lingered there after they had become extinct in most other parts of Britain. Isolation was no guarantee of survival, however, especially since local human residents, though not numerous, were themselves enthusiastic predators. Cumbrian wolves, bears, wild boars, and beavers had long since disappeared, along with their lowland conspecifics, and in 1892 one nostalgic naturalist lamented that even the memory of the wild cat (which resembles a particularly formidable tabby) "no longer survives among the venerable dalesmen whose grandfathers were the chief instruments of its extinction." He further claimed that badgers, which had been plentiful through most of the eighteenth century, had vanished by 1875; that the numbers of pine martens "have of late years greatly decreased"; that "within the last thirty years the Polecat has become very scarce in Lakeland"; and that otters were threatened "by those who surreptitiously trap these fine animals."[49]

Birds were also vulnerable to hunters, such as Sir Henry Vane, who had sold the manor of Wythburn to the Manchester Corporation. In its advocacy of the Thirlmere Scheme, the Corporation had proposed itself as the future protector of the lake's avian inhabitants. Once guns were banned from the reservoir watershed, it speculated, "what . . . should prevent the inaccessible crags from once more becoming the habitat of rare birds which have now all but disappeared from the district?"[50] Such optimistic prognostications were not, however, fulfilled by experience. Raptor populations, for example, did not immediately rebound at Thirlmere. The Corporation attributed this disappointing failure to local scofflaws rather than the disruptive preliminaries to construction or the shooting rights retained by Vane as part of the purchase agreement. During the period of his manorial lordship, John Grave contributed a series of letters on "The Wild Birds of Thirlmere" to the *Manchester City News.* In one of them he noted, in the language of outraged possession, that young falcons and buzzards had been taken from their nests to be sold in Keswick—"robbed from the rocks that are the property of the Corporation." Half a century later, the Waterworks Committee maintained a similarly vigilant (and inefficacious) stance, promising the local Association of Bird Watchers and Wardens that it would place newspaper advertisements and post notices "to the effect that persons trespassing on Corporation land for the purpose of (inter alia) taking wild birds' eggs will be prosecuted."[51]

The transformation of habitat proved at least as inimical to birds as overt predation, however, and so, despite its aggressive enforcement of the reservation boundaries, the Waterworks Committee was inevitably blamed for fluctuations in avian numbers. Thus in 1927 afforestation protesters made a corollary complaint about the absence of songbirds in the neighborhood of Thirlmere. The official response confirmed the worst anticipations of the critics. Corporation representatives admitted that evergreen monoculture did not ordinarily support the kinds of insects that such birds lived on but asserted that practical forestry priorities superceded any claims on behalf of wildlife. There were occasional exceptions, although these were motivated by pragmatic considerations, not by any acknowledgment of competing considerations. For example, several decades earlier, the woods had been infested by sawflies, relatives of bees and wasps whose hungry larvae had caused severe defoliation of the plantation conifers. To combat this insect plague, the reservation management had erected nesting boxes to lure the birds back. Once the sawflies had ceased to threaten the lumber crop, however, there was no further economic incentive to indulge the birds, and many of the boxes were harvested along with the surrounding trees. People who pined for birdsong, it was suggested, might be consoled by the fact that, at least for humans, "there are few pleasanter places than the woods . . . where the air is pungent with balsam and resin."[52]

But the Mancunian regime turned out to have ecological benefits as well as ecological costs. The evergreen plantations, although they supplanted native flora and failed to nourish songbirds, provided a refuge for some animals that were vanishing even from other parts of the Lake District. Writing to the *Lancashire Daily Post* in 1919, "Rusticus" called the Thirlmere reservation a "sanctuary for our rare wild things."[53] Foxes and badgers could live unpersecuted under the tree cover or on the uplands, and the woods also sheltered a small population of red deer, the largest nonextinct British mammal. (Although by the nineteenth century most red deer resident in Britain descended from animals reintroduced from the Continent, those of the Lake District had a plausible claim to represent native stock.)[54]

And despite the Corporation's general banishment of sheep from the Thirlmere watershed, the reservation ultimately also accommodated a historic flock of Herdwicks, a breed considered to have special connection to their natal territory. The flock had belonged to Isaac Thompson, who had farmed beside Thirlmere since before its conversion, and whose grandfathered holdings included the remnants of a number of smaller farms that had been mostly drowned by the expanded reservoir. His death in 1938 at the age of eighty first provoked local alarm that Manchester would not allow a new tenant to maintain any livestock. It was excitedly pointed out that sheep grazing high on Helvellyn were unlikely to contribute significant impurity to the reservoir, and that in any case, there was "a far greater danger of impurity to the water supply from the King's enemies than from farming stock."[55] Further, raising sheep would supplement the national food supply, a serious consideration in wartime. But securing a continued ovine presence did not allay all fears; the purity of the sheep themselves was also an issue. Acknowledging that "the question of crossing or keeping pure the Herdwick flock on the farm has been the subject of much anxious discussion," in 1940 the Waterworks Committee promised to maintain at least one thousand purebred Herdwicks.[56]

In addition to the desire to maintain the vestiges of traditional land use, concern about the fate of this flock reflected the Herdwicks' iconic local status. Although obviously domesticated and demonstrably nonnative, the breed had come to symbolize Lake District landscape. In part, this was because of the dearth of alternatives. Large mammals tend to be the most compelling representatives of nature, wild or otherwise, but the Cumbrian fauna was relatively impoverished, and what remained was hard to find. The sheep were certainly more conspicuous than the most distinctive feature of Lake District natural history: the alpine floral assemblage of the high fells, the residuum of the vegetation that colonized the entire region after the retreat of the last Pleistocene glaciation. It is unique in England, although similar to that of the high altitudes of Scotland and Wales (as well as to that of tundra areas much further north). Because of Helvellyn's elevation, within what was sometimes called the "infer-arctic zone," and because of the lime contained in its calcareous rock, such plants flourished on the heights above Thirlmere's eastern shore.[57] An early-twentieth-century geographical survey of the area noted that "in the belt between 1800 and 2700 feet we find a remarkable assemblage of plants of an alpine character, such as the kidney-leaved sorrel . . . , the rose-root . . . , the purple flowered saxifrage . . . , and the alpine rue."[58] Since no trees would grow at that altitude and in that soil, this distinctive vegetation was untroubled by encroaching evergreens, and Manchester's inclination to defend the reservoir from contamination by man and beast had the entirely incidental effect of protecting a rare and vulnerable floral community. Of course, the early critics of the Thirlmere Scheme or of the completed reservoir would not have considered this small triumph compensation for the larger defeats. They might not even have noticed it. The ecological perspective, which perceives each landscape as an intermeshed and interdependent association of plants and animals, was only beginning to emerge among scientists at that period.[59] It had not yet been integrated into public debate.

The Shadow of Things to Come, Part 1

The perspective represented by the original resistance to Manchester's grand water project, on the other hand, had become increasingly common. Any similarly transformative proposal was apt to encounter opposition from people who organized to defend landscapes that they neither inhabited nor owned. The confrontation pioneered by the Thirlmere Defence Association and the Manchester Corporation was to be replayed again and again, with preservationists striving to thwart what they viewed as the juggernaut of progress, until it became a routine component of the development process in any area that could be considered unspoiled. As had been the case with the TDA, these impassioned resistance efforts usually ended in failure. Nevertheless, they produced bad publicity, expense, and delay for developers, who were therefore anxious to minimize their impact. The initial debate over the Thirlmere Scheme, and Manchester's subsequent management of ongoing criticism, came to serve as a kind of template for those embroiled in similar struggles. In the first decade of the twentieth century, for example, the city of San Francisco targeted the spectacular Hetch Hetchy valley, over 150 miles distant in the Sierra Nevada mountains, as the site of its future water supply.[60] The fact that the valley had been protected as part of the newly designated Yosemite National Park inevitably heightened resistance to the plans to detach and drown it. After its completion, the San Francisco city engineer described the project in words that might also have been applied to the Thirlmere Scheme: "I never handled any proposition where the engineering problems were so simple and the political ones so complex."[61] To counter a very effective and energetic protest campaign, led by the Sierra Club and its founding president, the naturalist John Muir, the San Francisco planners turned to Manchester for advice.[62]

They sent John Freeman, the consultant engineer who was ultimately responsible for designing the Hetch Hetchy reservoir, to Europe in the summer of 1910 to examine the water supplies of other large urban areas. He paid special attention to those of the British cities that had undertaken ambitious reservoir and pipeline projects in the previous decades. But his focus was not only on their technical aspects. Freeman understood that the least tractable obstacles to the Hetch Hetchy project were likely to be political. So he laid the groundwork for arguments in the court of public opinion by emphasizing the beauty of these artificial, or artificially enhanced, lakes, as well as the recreational opportunities that they offered.[63] In Wales he admired Liverpool's Vyrnwy reservoir, which was "within a beautiful mountain region . . . and which has a great highway, patronized by the nature lovers, along its shores," as well as Birmingham's reservoirs in the Elan valley, where he found "their Craigh Gooch dam [to be] . . . the most beautiful waterworks dam in the world . . . a sight that any lover of beautiful cascades might well travel miles to see." He rode the steamboat across Glasgow's reservoir at Loch Katrine, enjoying the company of numerous fellow tourists. But it was Thirlmere—"in the very heart of the nature lovers' stamping ground"—that inspired his greatest enthusiasm. He inspected the reservoir in a hired car, noting with approval the new road from which he estimated that ten thousand visitors per year could view the lake, as well as the cesspools that handled the waste of nearby inns and stables, tidily engineered so that the pure reservoir water could flow unfiltered to urban consumers.[64]

Freeman also visited Manchester, where the secretary of the Waterworks Committee confirmed his favorable impressions. He recalled the early conflict and even showed Freeman a copy of *Extracts from the Lead-*

King's Head Hotel, 1910.

ing Journals on the Manchester Water Scheme, the collection of newspaper and magazine attacks that had been reprinted by the Thirlmere Defence Association. "Bishops and other prominent citizens had protested by all that was sacred against defiling and debasing this beautiful region," he acknowledged, but over time that had changed, and "everyone now admitted that the work . . . had been helpful to the nature lover."[65] Freeman was readily persuaded by this radically simplified account, and his own persuasive powers were instrumental in gaining federal approval for the Hetch Hetchy plan.[66] In his published report to the mayor and city attorney of San Francisco, a romantic image of Thirlmere by moonlight embellished the list of "foreign examples" of reservoirs that had enhanced the beauty and public enjoyment of the sites they had flooded. He rehearsed the story of how the "great popular outcry . . . in which bishops, baronets, actors and literary artists joined with great fervor," had ultimately been silenced by the benefits of the Thirlmere Scheme, and especially by the "beautiful macadam road circling the lake" which had become "one of the most popular holiday routes in England, . . . much traveled by coaching parties, automobiles and groups of cyclers."[67] He cited these examples "in order to prove the utter absurdity of the statement industriously circulated" by opponents of the Hetch Hetchy reservoir, that it "would exclude from its watershed tourists and campers, . . . or lessen the pleasure to be found within . . . the Yosemite reservation."[68]

The Shadow of Things to Come, Part 2

The city fathers of Manchester also had ample opportunity to benefit from their own experience. Large as it was, Thirlmere did not satisfy Manchester's growing thirst for long. In the twentieth century the city returned repeatedly to the Lake District to incorporate additional lakes into its water supply system. On the first occasion, the target was Haweswater, the highest and deepest of the lakes, which had been a runner-up to Thirlmere in the 1870s. The formal process of acquisition followed a similar pattern. In 1918 the Waterworks Committee warned the city council that the water supplied by Thirlmere and Longdendale combined would soon become inadequate and urged that an additional reservoir be constructed at Haweswater. The words of the committee chairman carried great weight, and the council acceded rapidly to this proposal. The Corporation immediately set the legislative wheels rolling, and parliamentary authorization for the project was granted in December 1919. As with the Thirlmere Scheme, the bill to enable the Haweswater Scheme encoun-

Antediluvian Haweswater.

tered resistance, but it was neither so vociferous as had been the campaign mounted by the Thirlmere Defence Association nor so effective.

The reservations about the proposed new reservoir echoed those expressed in the earlier debate, and they came from similar sources. (The major exception was local property owners, since the lake and watershed belonged to a single amenable individual.[69]) As before, some of the opposition came from within the Manchester city government. Officials who valued fiscal prudence were predictably reluctant to embark on such a large and expensive undertaking, especially if important decisions were made in a spirit of enthusiasm rather than deliberation. A future mayor criticized his fellow city councillors for endorsing the project after hearing only a brief speech by the chairman of the Waterworks Committee, without even taking the trouble to read the engineer's report.[70] Soon after the beginning of construction, it became clear that the committee's prediction of immediate shortage had been based on overly generous estimates of growth in both population and water usage, and so, a decade and a half later, another councillor (the wife of the earlier critic, as it happened) regretfully reported that she could "not feel that the Haweswater scheme is an example of far-sighted planning at its best."[71]

Some of the opposition came from municipalities that feared that

Manchester's appropriation of Haweswater would deprive them of a potential source of water. In response the parliamentary select committee appointed to consider the bill required that the Corporation provide water (at a reasonable rate, which meant without making a profit) to localities along the route of the pipeline from Haweswater to Manchester, and also to those near Manchester without the financial resources to find their own remote supplies.[72]

And some of the opposition came from people seeking to preserve the lake and its surroundings and to guarantee continued public access to them. Although the National Trust, which had been founded by some of the activists who had failed to save Thirlmere, was unable to prevent the sale to Manchester of property in Mardale (the valley in which Haweswater lay), defenders of the Lake District could console themselves with at least a small parliamentary victory. The original bill proposed to exclude walkers throughout the large watershed area (twenty-four thousand acres, more than double the size of the Thirlmere reservation), but as a result of the urging of the Commons, Open Spaces, and Footpaths Preservation Society, the final act mandated public access to much of the Haweswater reservation.[73]

Consolation was thus required because the proposed fate of Haweswater was similar to that which had already befallen Thirlmere. When Haweswater was dammed, its water level would be raised ninety-five feet and its area enlarged so that it flooded much of Mardale, submerging a church, an inn, and the homes and farms of its small population. From the perspective of the defenders of the Lake District, as well as that of the residents who would have to abandon their homes and livelihoods, this was a significant loss. In its original condition, one Victorian guidebook described Haweswater as "a fairy lake . . . very still and lonely, and . . . very beautiful—with fine breadths of grassy meadow land occasionally skirting its shore, and rich picturesque clumps of ancient forest, fantastic craggy points, and overhanging cliffs"; a little further up the valley the village of Mardale Green was "remarkable for its romantic position."[74] At the end of the valley was Mardale Head, "wild and solitary, the very *beau ideal* of romantic grandeur and sublimity."[75] Laments for the impending disappearance of this dramatic landscape were nevertheless a pale shadow of those that had been sparked by the threat to Thirlmere, often imbued not with outrage but with the gentler tones of nostalgia and resignation. A guidebook to "this wonderful corner of England" speculated that "possibly new beauties may emerge when the higher level of the water materialises, but certainly for a time . . . the dale will be in a state of making and its matured charms will belong to a period definitely closed." A lover of sheep complained that, "to Manchester, mutton and wool on the mountains are far less important than the possibility of capturing another inch of rain per annum from closely afforested slopes." A frequent visitor, observing the valley when the works were already in progress, conceded that although "there is little romance or sentiment about a broad new highway," it was "hard to visualise Haweswater of the future."[76]

Perhaps one reason that resistance to the impending transformation of Haweswater was more subdued than that which had greeted the Thirlmere Scheme was the very peace and seclusion celebrated in travel accounts.[77] Whereas Thirlmere lay beside the main road between Keswick and Kendal, Haweswater was (and remains, except to fell walkers) accessible only via a minor road that terminates in the valley. Its physical charms were unenhanced by the literary glamor of many other Cumbrian lakes and valleys. Although Thornthwaite Hall, the setting of *Can You Forgive Her?*, the first novel of the Palliser series, lay nearby, Trollopian associations were no match for those of Wordsworth and Coleridge.[78] Most tourists were content to appreciate the dramatic Lake District landscape in more accessible or more celebrated spots; as a later chronicler put it, "only the discerning visitor penetrated into the valley."[79] Or perhaps the example of Thirlmere—and of other minor victories and major defeats that had followed it—had instilled a sobering sense of possibilities and probabilities.

Once Manchester's plans became public, however, the anticipated conversion of the lake prompted a certain amount of commentary, most of which rehearsed positions that had been articulated with regard to Thirlmere. Both antagonists and proponents often included explicit comparisons with the previous scheme, even though, almost four decades after the start of construction, few of them could have retained clear memories of the undammed lake. Thus a journalist who characterized Haweswater as "the next victim of the octopus of civilization" recalled how the "wild beauties" of Thirlmere had been "tamed by the Manchester Water Scheme authorities," while the author of a letter to the editor of the *Times* more moderately hoped that "the Manchester Corporation will have profited by public criticism, and not repeat architectural incongruities with which the roadside near Thirlmere . . . has been disfigured."[80] One Lake District newspaper warned that Manchester might attempt to evade its responsibility to provide other towns and rural areas with water, while to a Manchester newspaper it seemed "a pity [that] . . . water supply cannot be secured without expensive litigation and opposition."[81]

Advocates of the new scheme suggested that Thirlmere had benefited from its transformation, although not everyone would have been encouraged by the terms of their praise; for example: "Some old landmarks were naturally submerged, but . . . Thirlmere has gained rather than lost in attraction, whilst the owners of wheeled vehicles . . . have great good reason to bless the Manchester Corporation."[82] Similarly, one city councillor, a frequent visitor to Mardale who valued its "peaceful seclusion" as an escape from "the hurry and scurry, the noise, discomfort and danger of charas, motor cars and motor cycles rushing . . . along the main roads of Lakeland," predicted that "the present generation will lose . . . the valley and all it contains" but nevertheless concluded that "none of its beauty will be lost."[83] And resisters would surely have been dismayed by the bland assurances offered by the chairman of the Waterworks Committee at the jovial dinner concluding a visit of the entire Manchester Corporation to the Haweswater site, although his immediate audience received them enthusiastically. He predicted that "there will be no more charming spot in the country than Haweswater when we have finished with it. (Hear, hear.) No one I hope has an idea . . . that we have any intention of spoiling the beauty of the country. (Applause.) It is our intention to improve the appearance of the whole surroundings—(hear, hear)—because we shall do as we have done at Thirlmere. We shall begin afforestation as soon as we can."[84]

It thus seemed unlikely that the Waterworks Committee had profited from public criticism of the Thirlmere Scheme—or, at any rate, that it approached such projects differently as a result. On the contrary, the completion of dam and pipeline had been triumphal events for its members, confirming their initial opinions, and their successors had inherited their attitudes. Even the neo-Gothic style that was alleged to "disfigure" the buildings at Thirlmere was repeated in the draw-off tower at Haweswater, although not in the alien red sandstone that had offended some obervers. The tower, like the hotel that Manchester also built, was made of gray local stone, much of it recycled from the demolished Mardale church.[85] The Thirlmere events, of course, had simultaneously confirmed the opposite opinions of the Scheme's most committed critics. And in the case of Haweswater, those opposed to reshaping the Lake District to fill urban needs showed no greater inclination to alter their positions. Much of the debate about the project occurred in a kind of vacuum, because construction did not begin until 1929, ten years after the original approval, and even then progress was slow. But once there was something to look at, critics responded in predictable style, invoking the rhetoric of desecration and sin, as well as that of aesthetics and class. For example, H. H. Symonds, who liked automobiles as little as he liked conifer plantations, called the new road that extended along one side of the reservoir "a real disaster, an engineer's vulgarism . . . sheer wickedness, an otiose parade ground for those who sit in cars and wax fat. . . . One is glad to think how much interest, and for how long, the ratepayers are paying on the money borrowed to build it."[86]

Looking Backward

Haweswater took a long time to build, as well as to pay for. After its delayed start, construction was interrupted by the economic depression of 1931, and water from the new reservoir did not reach Manchester until 1941. At first Haweswater supplies were conveyed through the Thirlmere aqueducts, but a direct pipeline from Haweswater to Manchester, completed in 1955, increased its output.[87] Schemes for drawing still more water from the reservoir were mooted in the 1970s, as demand increased throughout northwest England, but they remained in the realm of suggestion.

In addition to threatened landscape, objections to heightening the Haweswater dam or building subsidiary dams often invoked the fate of Mardale Green, the tiny village whose incipient demise had not figured prominently in the original resistance to the reservoir. But although dead—and in a sense more than dead, because even the remains in its graveyard had been exhumed and relocated—Mardale Green refused to stay buried. Symbolically, since its submergence, it had come to represent the human cost of reservoir creation, an increasingly prominent component of the critiques of all major environmental transformations. And literally, during dry summers, it (or what was left of it) appeared again. These ghostly reemergences became popular attractions, drawing television crews and reporters from the national newspapers, as well as hordes of sentimental tourists with their attendant purveyors of souvenirs and refreshments. In the summer of 1984 cars and vans converged on the reservoir in such numbers that police had to close the road into the valley.[88]

Haweswater was not Manchester's only target in its quest for additional Cumbrian water. the city subsequently managed to tap several of the largest lakes. But changes in popular attitudes and political realities made these accessions much more challenging and complex. A 1961 bill that would have enabled its initial attempt to draw water from Ullswater and to dam a nearby valley failed in Parliament, having aroused opposition from the widest possible range of stakeholders, from rock climbers and

Haweswater flooded.

Haweswater dried.

landscape preservationists to the Cumberland County Council and the Cumberland River Board.[89] Critics often recalled Manchester's acquisition of Thirlmere and Haweswater in the language of invasion and plunder, and their heated accusations gained force and credibility from the Corporation's tactless repetition of methods that had previously proved inflammatory. The projectors still did not take the trouble to explain their proposals to potentially interested parties in advance. Worse yet, local residents discovered Manchester water engineers covertly exploring the Ullswater neighborhood in advance of any public announcement of their intentions, evoking the stealthy surveillance that had preceded the launch of the Thirlmere Scheme.[90]

This check on the Corporation's plans could not be permanent: Manchester's water problem remained to be solved. So in 1965 Parliament authorized a very different proposal, which allowed the city to abstract water from Ullswater and Windermere, via underground pumps that would neither alter the scenery nor interfere with the activities of visitors to the lakes and their surroundings. In contrast to the near-absolute control that the Corporation exerted over the large reservations surrounding Thirlmere and Haweswater, at Ullswater and Windermere it was granted access only to the water. And this water was very different from that of the two existing reservoirs. Windermere, the largest natural lake in England, was the most intensively developed lake in the Lake District, and both lakes had long histories of residential and commercial activity. For the first time, however, Manchester was unable to purchase a buffer zone around the lakes and barred from interfering with their established uses, which meant that the pure waters of Thirlmere and Haweswater would mingle with water tainted by human occupation and industry. After almost a century of drinking Lake District water as it had fallen from the sky, Manchester consumers had to fund a water treatment facility.[91]

Public opinion was not the only thing in flux after World War II. The Labor government that swept to power in 1945, soon after the end of the European conflict, put the nationalization of services and industries high on its agenda. Coal, gas, electricity, and the railway system were nationalized by the end of the decade, but water supply, although an obvious target for higher-level administration, remained under local control. The complexity of the existing pattern of ownership made reform difficult, and local authorities, including the municipalities from which the Labor Party drew much of its electoral support, were reluctant to relinquish possession. Much later, when the tide of nationalization was on the ebb, the Water Act of 1973 established regional water authorities, with responsibility for sewage treatment, pollution control, and conservation, as well as for water supply.[92] The Manchester Corporation then ceded its elaborate system of reservoirs and pipelines to the North West Water Authority, which in 1989 was privatized (this time on the crest of the wave) as North West Water. In 1995 it merged with Norweb (the privatized version of the former Northwestern Electricity Board) to form United Utilities, which now manages water supplies not only in various parts of Britain, but in Australia, Bulgaria, Canada, Estonia, India, the Philippines, and Poland.[93]

The management of the Lake District itself also changed significantly during this period. The National Trust, established in 1895, was an outgrowth of the movement to preserve access to open spaces, but instead of campaigning to persuade others to maintain scenic landscapes and historic buildings, it proposed to acquire them itself.[94] As Rawnsley put it, the Trust would ensure that such places remained "safe for the enjoyment of the people, unharmed by the speculating builder and free from the restrictions which land preserved for sporting rights must necessarily involve." Lake District properties were among its earliest purchases, and over the years it accumulated much more Cumbrian territory than the Manchester Corporation ever owned.[95] Piecemeal preservation had inherent limitations, however, no matter how large the pieces ultimately grew. Preservation of the entire area was an old idea: Wordsworth had concluded his *Guide to the Lakes* by calling "the district a sort of national property, in which every man has a right and interest who has an eye to perceive and a heart to enjoy."[96] It was also an idea whose time was slow in coming. In the 1930s, the National Trust joined with a number of other organizations to urge the creation of a national park in the Lake District. The National Parks and Access to the Countryside Act established the legal basis for realizing these long cherished hopes in 1949, and the Lake District was designated as a national park in 1951.[97] The park owns only a small percentage of the land within its boundaries, but it regulates land use in an attempt to ensure that the area is maintained and developed in accordance with its goals of conservation and recreation.

For the centenary of the Longdendale reservoirs in 1947, the Manchester City Council published a booklet commemorating all the achievements of the Waterworks Committee. Predictably enough, it presented the "provision . . . of an adequate and pure supply of water" to "one of the most densely populated industrial areas of the country" as "a notable tribute to the foresight, energy and courage of a succession of public representatives during the past 100 years." Ignoring the protests that had persisted over the years, it dismissed aesthetic objections to the Thirlmere Scheme, praised the sideline in commercial forestry, and stated proudly that "the policy of the Waterworks Committee has been to purchase whenever possible the fee simple of the gathering grounds in order to ensure purity of the water by controlling and restricting the agricultural and other activities on the land."[98] By the centenary of the reservoir at Thirlmere, however, the Waterworks Committee no longer existed because the city no longer possessed any waterworks. Its water was supplied by a private company. And changing environmental views had radically altered the priorities that guided rural development. Nevertheless, Manchester celebrated the later anniversary in terms that highlighted continuity, augmented by such modern touches as jazz bands and street theater in front of the town hall, and a convoy of vintage vehicles that traced, as nearly as possible, the route of the aqueduct (led by a tanker truck full of Thirlmere water).[99]

Echoing the elaborate festivities that had heralded the Scheme's completion, the mayor of Manchester traveled to the reservoir to raise a glass of "pure Thirlmere water" with the water treatment manager, the North West Water operations director, and a former Manchester Corporation forester.[100] The operations director unveiled a small metal plaque, set modestly beside the imposing granite monument of the previous century. Adorned by the soon to be obsolete North West Water logo, it stated simply, "1894–1994. To celebrate the centenary of the Thirlmere Water Supply Scheme." After the tanker arrived in front of the town hall, the mayor, who had beaten the convoy back to the city, hosted a grand civic reception.

The congratulatory spirit spread to other towns that benefited from the reservoir. A Wigan journalist admonished readers, "next time you brew up, or leap into the bath, [to] spare a thought for the system that brought you that clear running water," while a Bolton newspaper claimed that "much of the pioneering thinking and engineering expertise that led to a clean and regular water supply . . . blossomed in Bolton."[101] Coverage of the event followed a similar pattern, whether in national newspapers or local ones, in dailies or in weeklies. Tucked into the middle of most articles was an acknowledgment of the early resistance, which was then dismissed as misguided in either its judgment (there was no damage to the landscape) or its priorities (the basic needs of the many outweighed the predilections of the few). Picking up the cudgels of the original debaters, one Manchester newspaper even asserted that the heavily forested road around Thirlmere was "one of the most beautiful drives in the Lake District."[102] North West Water reproduced much of this favorable coverage in a booklet. It also provided a more substantial commemoration, in the form of a paperback prefaced by the chairman of the company, who recalled the "delighted

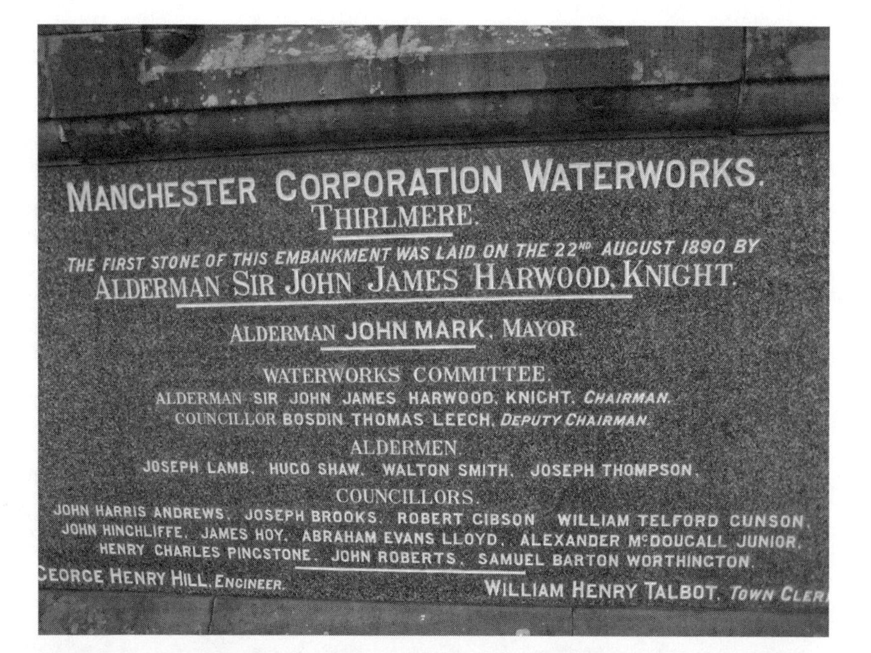

The granite monument commemorating the beginning of work on the dam and the more restrained plaque marking the centenary of its completion.

Thirlmere in 2007, looking south from the embankment.

crowd" that had witnessed the arrival of the first Thirlmere water in 1894.[103] After rehearsing the history of the Scheme, the authors concluded, "It is quite clear now that the eighty years or so of distinctive measures adopted by Manchester have resulted in the preservation of one of the few lakeside areas of unspoilt quiet in the Lake District."[104]

If the view from Manchester was thus resolutely triumphalist, there continued to be another side to the Thirlmere story. Aficionados of Lake District scenery found it nondescript—merely pleasant at best. Although its position beside the A591 made it, willy-nilly, part of the itinerary of most visitors to the region, it remained, as it had been since guidebooks first were written, a secondary attraction. And whatever its appeal, the reservation continued to be difficult to enjoy, since, as one guide book put it, only vigorous and contrary-minded walkers would not be "deterred by the many TRESPASSERS WILL BE PROSECUTED signs that Manchester Corporation have sprinkled about the landscape."[105] The North West Water Authority followed in Manchester's vigilant footsteps, even though the construction of the water treatment facility after Ullswater and Windermere joined the system had eliminated the need to defend Thirlmere from touristic contamination. In 1989 the environment secretary pointedly chose Thirlmere as the site for a speech about the right to roam; when he proposed legislation to ensure that the privatization of water supply would not result in further restriction of public access, Thirlmere and Haweswater were two of only three locations singled out as in need of special guarantees.[106] Perhaps in response to public concern about access, or to a spate of bad publicity following water shortages in drought years, along with reports of asset stripping and inadequate maintenance, the environs of the reservoir have become more welcoming. There are now small parking lots and marked trailheads scattered along the western side of the lake, along with a jetty for rowboats. Icons on the United Utilities website promise visitors climbing, fishing, water sports, a bridle path, a viewpoint, and an environmental trail.[107] Dalehead Hall has become a hotel. But the expanded and depopulated lake is indeed very still, except for the muted hum of road traffic and, at certain seasons, the buzz and crash of the timber harvest.

Epilogue

All modern environmental arguments rest on predictions—usually (although not always) about benefit to some people or about harm to landscape, flora, fauna, and other people. Like all predictions, they are, by definition, unprovable. That is to say, they are unprovable at the time that they are made. The controversy surrounding the planning and construction of the reservoir at Thirlmere was filled with such predictions, and the lapse of more than a century has made it possible to assess them. Somewhat disconcertingly, it turns out that most of them were correct—both the positive and the negative. The water from Thirlmere did improve the life of ordinary Mancunians and enrich the fortunes of the city's elites. The expense of the enormous project was well justified in terms of consumption; indeed the supply proved inadequate to the city's needs within a few decades of the dam's construction. The thinly settled community of the Vale of Thirlmere and St. John's Vale was significantly diminished when Manchester raised the water level and prohibited most economic activity within the lake's watershed. And the lake was irretrievably transformed, with aesthetic, social, and environmental consequences about which no consensus has ever emerged.

So, at least with regard to evaluation, hindsight seems less clear in this case than it is often reputed to be. Perhaps one reason for this persistent ambiguity is that the terms of the controversy have not changed much over the intervening period, although the versions in which they were deployed by the TDA and the Manchester Waterworks Committee may seem a little crude to modern ears. But they are not so crude as to be unrecognizable—the rhetoric of preservation and progress (or of reaction and spoliation, to put them in negative terms) continue to define opposed and often irreconcilable positions in environmental confrontations. Another reason that it is difficult to evaluate the results of the Thirlmere controversy, even at this chronological remove, is that the assessment of any course of action must depend on the alternatives and on the contexts. Even in the 1870s, it was clear that the future held some kind of development for every reasonably large and accessible Cumbrian lake. Windermere and Derwent Water were already suffering from the attentions of trippers. Their subsequent history, along with that of Ullswater and Grasmere, among others, suggest that Thirlmere was not alone in its transformation, although it has been altered by different forces and in a different cause. Indeed, the jealous care of the Manchester Corporation and its successors may have produced a result that the lake's original defenders would have preferred to routine commercial development.

The basic structure of the Thirlmere debate has been replicated in controversies about many other settings threatened with similar transformation during the last century and a quarter, although the political landscape has shifted dramatically and scientific understanding of environmental issues has deepened. In most places, official regulation plays a more significant role in shaping large-scale developments than it did in Victorian Britain. A few new voices have joined the discussion, expressing perspectives that were either muted or missing in the original controversy—most notably those of expropriated residents (especially small farmers and laborers) and those of ecologists. The sense of entitlement on the part of remote stakeholders has become increasingly global in its reach and in the penetration of its arguments. Powerful as these additional interventions can be—politically, intellectually, and emotionally—they have had little influence on the formulation of the core opposing positions, with respect either to the continuing skirmishes at Thirlmere or to analogous controversies that have arisen throughout the world. The conclusion of the struggle over Hetch Hetchy would have seemed familiar to participants in the Thirlmere controversy, as would the similar conclusions with regard to such later and larger projects as the Aswan Dam and the Three Gorges Dam.

This consistency has posed particular problems for preservationists, because they have been the most frequent losers. Like reenactors on the Cavalier side in the English Civil War, or the Confederate side in the American one, they have become accustomed to noble defeats. Yet they still seem reluctant to address some of the issues that tripped up the TDA. Acknowledging that both Thirlmere in particular and the Lake District as a whole had pasts that were checkered and volatile, rather than static and pure, would have confronted the lake's defenders with a series of difficult questions. What are the indications that preservation is appropriate and necessary? What is the ideal past condition at which restoration should aim? If some changes have taken place in the past, on what grounds should other changes be prevented in the future? Such notions as "pristine wilderness" and "unspoiled countryside" are emotionally powerful, but the claims inherent in those appealing and evocative labels are often belied by evidence provided by the threatened landscapes themselves. Of course it is more difficult to answer a series of questions to which the answers are likely to be complex and ambiguous than it is to stand by a slogan or a credo. Absolute positions are more compelling than nuanced or intermediate ones—easier to articulate though more difficult to defend. Every individual's convictions are his or her own business, and strong commitments can be admirable. But in the realm of political action they can also be dangerously misleading, inclining the people who hold them to equate the strength of their position with the strength of their feelings. The defenders of Thirlmere made a big splash, but they never stood a chance, and the same was true of the defenders of the Hetch Hetchy Valley. The same is still too often true of environmental advocates, and of the causes that they champion.

Acknowledgments

Many people and institutions have given me help and encouragement during the years that I have worked on this book. My research took me to a number of collections in Britain and the United States. I owe thanks to the librarians and archivists at the Manchester Central Library (especially the Archives and Local Studies department), the Greater Manchester County Record Office, the Cumbria County Record Offices at Carlisle and Kendal, the British Library, the Armitt Library, the Institute of Civil Engineers, the National Humanities Center, and the university libraries of Harvard, Cambridge, and the Massachusetts Institute of Technology. I am especially grateful to Drew Laughland of the GIS Services at the MIT Libraries for producing the maps. A John D. and Catherine T. MacArthur Foundation Fellowship from the National Humanities Center and leaves from the Massachusetts Institute of Technology have given me time for writing and research. Several excellent UROP (Undergraduate Research Opportunities Program) students from MIT and Wellesley have helped with my research: Michelle Duvall, Cortney Harding, Sam Hollander,

Aysa Kaya, Wendy Kim, Doug Kriner, Fenny Lin, Christine McEvilly, and Jennifer Meanwell. And I have benefited greatly from the expert assistance of two graduate students: Rebecca Woods and Andrew David. It has been a pleasure to work with my editor Christie Henry at the University of Chicago Press as I completed the manuscript. Mabel Chin and Margo Collett of the History Department office at MIT have offered essential and unfailing support.

Like most large projects, this one has existed in a number of preliminary pieces and versions. I am grateful to many people who offered comments on draft chapters and other invaluable advice and support, including Gillian Beer, Eric Caines, Karen Caines, Nancy Cott, Allen David, Catherine David, Deborah Fitzgerald, Ginger Frost, John Kucich, Michele Lamont, Peter Mandler, David Matless, Anne McCants, Craig Murphy, Jon Parry, John Pickstone, Leah Price, Steve Pyne, Jonathan Riley, Molly Rothenberg, Diane Sadoff, Anne Secord, Jim Secord, Mart Stewart, Jay Taylor, Andrew Warren, Jennifer Warren, and JoAnne Yates—and especially to the graduate students in the PXY group: Etienne Benson, Nick Buchanan, Shane Hamilton, Jenny Smith, Bill Turkel, and Anya Zilberstein. I have also appreciated the responses of audiences at Bard College, the University of British Columbia, the University of California at Santa Barbara, the University of California at Santa Cruz, Cambridge University, Dartmouth College, the Dibner Institute, Harvard University, Kalamazoo College, the University of Leeds, the University of Manchester, the National Humanities Center, the North American Conference on British Studies, the North American Victorian Studies Association, the University of North Carolina at Charlotte, North Carolina State University, the University of Reykjavik, Salem State University, Stanford University, Western Washington University, and Yale University.

Notes

Chapter 1

1. John James Harwood, *History and Description of the Thirlmere Scheme* (Manchester: Henry Blacklock, 1895), 27–28.

2. Roby X., "Thirlmere, Past and Present. Part II," *North Lonsdale Magazine and Furness Miscellany* 4 (October 1900), 54.

3. Harwood, *History and Description,* 60; Ian Tyler, *Thirlmere Mines and the Drowning of the Valley* (Keswick: Blue Rock Publications, 1999), 143.

4. Tyler, *Thirlmere Mines,* 140.

5. The accusation came from E. S. Howard, a member of Parliament from Cumberland, in the debate on the second reading of the Manchester Corporation Water Bill of 1878. *Hansard Parliamentary Debates,* 3rd ser., vol. 237 (1878), col. 1508.

6. "Instructions for Cross Examination of Mr. Alderman Grave," prepared for Harwards, Shepherd, and Mills (Manchester Central Library, M48, box 11).

7. Henry Irwin Jenkinson, *Jenkinson's Practical Guide to the English Lake District* (London: E. Stanford, 1872). This first edition was followed by a series of revisions that kept the guide up to date until Jenkinson's death in 1891.

8. Harwood, *History and Description,* 43–44; "The New Manchester Waterworks: The Thirlmere Supply," *Manchester Weekly Times* (January 14, 1894).

9. Robert Somervell, *Chapters of Autobiography* (London: Faber and Faber, 1935), 51–52.

10. Daniel Defoe, *A Tour through the Whole Island of Great Britain,* ed. P. N. Furband and W. R. Owens (1724–26; New Haven: Yale University Press, 1991), 291. For an account of a slightly earlier tour (1698), see Celia Fiennes, *The Journeys of Celia Fiennes,* ed. Christopher Morris (London: Cresset Press, 1949).

11. Thomas Gray, *Thomas Gray's Journal of His Visit to the Lake District in October 1769,* ed. William Roberts (Liverpool: Liverpool University Press, 2001).

12. C. M. L. Bouch and G. P. Jones, *A Short Economic and Social History of the Lake Counties, 1500–1830* (Manchester: Manchester University Press, 1961), 279.

13. Herman Prior, *Guide to the Lake District of England,* 3rd ed. (Windermere: J. Garnett, n.d. [1870?]), 34.

14. John Bailey and George Culley, *General View of the Agriculture of Northumberland, Cumberland and Westmorland* (1805; Newcastle-upon-Tyne: Frank Graham, 1972), 204.

15. For a sense of the accommodations available to contemporary travelers, see Dave Walker and Kerry Walker, *Wordsworth and Coleridge: Tour of the Lake District, 1799* (Blackpool: David Walker, 1997), xiii–xvi.

16. Nikolaus Pevsner, *Cumberland and Westmorland* (Harmondsworth, Middlesex: Penguin, 1967), 37.

17. *A Picturesque Tour of the English Lakes, Containing a Description of the Most Romantic Scenery of Cumberland, Westmoreland, and Lancashire* (London: R. Ackermann, 1821), iii–v.

18. [Joseph Budworth/Joseph Palmer], *A Fortnight's Ramble to the Lakes in Westmoreland, Lancashire, and Cumberland. By a Rambler* (London: Hookham and Carpenter, 1792), xiv.

19. "Westmorland," *Quarterly Review* 122 (1867): 373.

20. On the relation between Wordsworth's work and nineteenth-century tourism, see James Buzard, *The Beaten Track: European Tourism, Literature and the Ways to "Culture," 1800–1918* (Oxford: Oxford University Press, 1993), 19–32.

21. Kenneth R. Johnston, *The Hidden Wordsworth: Poet, Lover, Rebel, Spy* (New York: W. W. Norton, 1998), chaps. 1–3.

22. House of Commons Select Committee on the Manchester Corporation Water Bill (Shorthand Writer's Notes), Minutes of Evidence, testimony of John Harward (March 14, 1878), 291.

23. William Wordsworth, *A Guide through the District of the Lakes in the North of England,* ed. Ernest de Selincourt (1835; Oxford: Oxford University Press, 1906), 1.

24. Thomas De Quincey, *Literary Reminiscences from the Autobiography of an English Opium Eater,* in *The Works of Thomas De Quincey* (Boston: Houghton Mifflin, 1851), 3:310–11.

25. Gray, *Journal,* 87.

26. Thomas West, *A Guide to the Lakes in Cumberland, Westmorland and Lancashire* (1784; Oxford: Woodstock Books, 1989), 82–83.

27. James Clarke, *A Survey of the Lakes of Cumberland, Westmorland, and Lancashire* (London: James Clarke, 1787), 118.

28. William Gilpin, *Observations Relative Chiefly to Picturesque Beauty, Made in the Year 1772, on Several Parts of England; Particularly the Mountains, and Lakes of Cumberland and Westmoreland* (1786; Poole, NY: Woodstock Books, 1996), 1:170–71.

29. [Adam Walker], *A Tour from London to the Lakes . . . Made in the Summer of 1791. By a Gentleman* (London: John Abraham, 1792), 107–8.

30. Ann Radcliffe, *A Journey made in the Summer of 1794 . . . to which are added, Observations during a tour to the Lakes of Lancashire, Westmoreland, and Cumberland* (Dublin: William Porter, 1795), 469–70.

31. William Green, *A Description of Sixty Studies from Nature* (London: Longman, Hurst, Rees, and Orme, 1810), 54–55.

32. Samuel Taylor Coleridge, Notebook, October 23, 1803, quoted in Grevel Lindop, *A Literary Guide to the Lake District* (London: Chatto and Windus, 1993), 119.

33. Edward Baines, *A Companion to the Lakes of Cumberland, Westmoreland, and Lancashire* (London: Simpkin and Marshall, 1834), 117.

34. "Wythburn Lake," *Housekeeper's Magazine, and Family Economist* (1826), 307.

35. Harriet Martineau, *Complete Guide to the English Lakes* (Windermere: J. Garnett, 1855), 71.

36. H. D. Rawnsley, *Literary Associations of the English Lakes* (Glasgow: James MacLehose, 1894), 2:218–24.

37. Quoted in Roby X., "Thirlmere, Past and Present. Part I," *North Lonsdale Magazine and Furness Miscellany* 4 (August 1900), 27.

38. "A Visit to the Lakes," *Ainsworth's Magazine* 18 (July 1850), 144.

39. Thomas George Bonney and Elijah Walton, *English Lake Scenery* (London: W. M. Thompson, 1875), caption to illustration 20.

40. October 20, 1877. Quoted in Thirlmere Defence Association, *Extracts from the Leading Journals on the Manchester Water Scheme* (Windermere: J. Garnett, 1878), 3.

41. J. A. Hassan and E. R. Wilson, "The Longdendale Water Scheme 1848–1884," *Industrial Archaeology* 14 (1979): 102.

42. For accounts of the engineering issues raised by the Longdendale Scheme, see Hassan and Wilson, "Longdendale Water Scheme," 102–21, and Tom Quayle, *Reservoirs in the Hills: The Story of the Construction of the Reservoirs in the Longdendale Valley Which Feed Manchester with Water* (N.p.: Senior Publications, 1988).

43. For Bateman's assessment of possible alternative sources, see John Frederic LaTrobe Bateman, *Reports on the Various Schemes for Supplying Glasgow with Water* (Manchester: Cave and Sever, 1853), 1–35.

44. Bill Gow, *The Swirl of the Pipes: A History of Water and Sewerage in Strathclyde* (Glasgow: Strathclyde Regional Council, 1996), 25–31; *Notes on the Water Supply of Glasgow, Prepared on the Occasion of the Celebration of the Jubilee of the Loch Katrine Water-Works, 14th October, 1909* (Glasgow: Glasgow Corporation Water Department, 1909), 18–45.

45. Later research on the science of water flow showed that the Admiralty's fears about the effect of the Loch Katrine Scheme on the Forth estuary were unfounded. Twentieth-century public health statistics showed that Loch Katrine water was indeed associated with lead poisoning, but at the time this concern was dispelled by the testimony of rival chemists retained by the city council. T. C. Smout, *Nature Contested: Environmental History in Scotland and Northern England since 1600* (Edinburgh: Edinburgh University Press, 2000), 104. For earlier accounts of the objections to the Loch Katrine Scheme, see *Notes on the Water Supply of Glasgow,* 23–32; John S. Clarke, *An Epic of Municipalisation: The Story of Glasgow's Loch Katrine Water Supply* (Glasgow: Forward, 1928), 16–27; and "The Glasgow Loch Katrine Water Scheme," *Glasgow Herald* (February 20 and February 24, 1854).

46. Ouida [Louise de la Ramée], "The Ugliness of Modern Life," in *Critical Studies* (New York: Cassell, [1900?]), 231. I am grateful to Kate Flint for bringing this to my attention.

47. *Punch,* October 29, 1859, reprinted in Nicholas Dickson, ed., *The Queen's Guard of Honour, Loch Katrine, 1859. A Memorial Record* (Glasgow: N. Macphail, 1885), 45.

48. At the turn of the century the *Rob Roy* was replaced by the SS *Sir Walter Scott. SS Sir Walter Scott on Loch Katrine* (Norwich: Jarrold Publishing, 1994), n.p.

49. *Ferguson's Tourists' Guide to Callander, the Trosachs, Loch Katrine, Loch Lomond . . . ,* 9th ed. (Glasgow: Thomas Murray, 1911), 25.

50. These were subsequently republished in a pamphlet entitled *Kendal and Windermere Railway,* reprinted as the appendix to William Wordsworth, *The Illustrated Wordsworth's Guide to the Lakes,* ed. Peter Bicknell (New York: Congden and Weed, 1984). On the coming of the railroad, see William Rollinson, *A History of Man in the Lake District* (London: J. M. Dent, 1967), 139.

51. J. D. Marshall and John K. Walton, *The Lake Counties from 1830 to the Mid-twentieth Century: A Study in Regional Change* (Manchester: Manchester University Press, 1981), 205–6; Jeffrey Richards, "The Role of the Railways," in *Ruskin and Environment: The Storm Cloud of the Nineteenth Century,* ed. Michael Wheeler (Manchester: Manchester University Press, 1995), 125.

52. E. L. Blanchard, *Adams's Pocket Guide to the Lake District of Lancashire, Westmoreland, and Cumberland* (London: W. J. Adams, 1852), iii; Geoffrey Berry and Geoffrey Beard, *The Lake District: A Century of Conservation* (Edinburgh: John Bartholomew, 1980), 1–2.

53. Rollinson, *History of Man,* 140.

54. Somervell, *Chapters of Autobiography,* 51. For a discussion of the resistance to the railroad and the later Thirlmere controversy in the context of developing environmentalist ideas, see Charles-François Mathis, *In Nature We Trust: Les paysages anglais à l'ère industrielle* (Paris: Presses Universitaires Paris-Sorbonne, 2009).

55. Somervell, *Chapters of Autobiography,* 51.

56. Robert Somervell, *A Protest against the Extension of Railways in the Lake District* (Windermere: J. Garnett, 1876).

57. Marshall and Walton, *Lake Counties,* 209–11.

58. *Saturday Review* (August 25, 1877); *Builder* (December 1, 1877). Quoted in Thirlmere Defence Association, *Extracts from the Leading Journals,* 11, 14.

59. J. Clifton Ward, "Are We to Preserve Our English Lakes?" *Daily News* (June 12, 1877).

60. John Edward Marr, *The Geology of the Lake District, and the Scenery as Influenced by Geological Structure* (Cambridge: Cambridge University Press, 1916), 7, 2; Edward L. Hull, "The Geology," in W. G. Collingwood, *The Lake Counties* (New York: E. P. Dutton, 1902), 243.

61. Marr, *Geology of the Lake District,* 1; Jonathan Otley, *A Concise Description of the English Lakes and Adjacent Mountains* (Keswick: Jonathan Otley, 1825), iii.

62. James Clifton Ward, *The Geology of the Northern Part of the English Lake District* (London: HMSO, 1876), 101.

63. Marr, *Geology of the Lake District,* 129.

64. Nathaniel Beardmore, *Manual of Hydrology* (London: Waterlow and Sons, 1862), 298–99; John Fletcher Miller, "On the Meteorology of the Lake District of Cumber-

land and Westmoreland, Part I," *Philosophical Transactions of the Royal Society of London* (1849): 88–89.

65. Hull, "Geology," 248. Although Victorian geologists concurred in recognizing the importance of glaciation in the Lake District, consensus on many important points did not emerge until the twentieth century. For an account of these debates, see David Oldroyd, "Early Ideas about Glaciation in the English Lake District: The Problem of Making Sense of Glaciation in a Glaciated Region," *Annals of Science* 56 (1999): 175–203.

66. William Hodgson, *Flora of Cumberland* (Carlisle: W. Meals, 1898), x–xii.

67. John Edward Marr, *Cumberland* (Cambridge: Cambridge University Press, 1910), 61.

68. H. A. Macpherson, *A Vertebrate Fauna of Lakeland, Including Cumberland and Wesmorland with Lancashire North of the Sands* (Edinburgh: David Douglas, 1892), 1–87; John Webster, "Mammals in Cumbria—A Centenary Review," in David J. Clarke and Stephen M. Hewitt, eds., *Cumbrian Wildlife in the Twentieth Century,* Transactions of the Carlisle Natural History Society 12 (Carlisle: Carlisle Natural History Society, 1996), 77–88.

69. Wordsworth, *Illustrated Wordsworth's Guide,* 105.

70. John Ruskin, *A Tour to the Lakes in Cumberland: John Ruskin's Diary for 1830,* ed. James S. Dearden (Aldershot: Scolar Press, 1990), 41.

71. Petition of Stanley Hughes Le Fleming of Rydal Hall, in "Thirlmere. Petitions against Bill. 1878" (Manchester Central Library, M231/2, box 2).

72. Bruce L. Thompson, "The Leathes Family of Dalehead," *Transactions of the Cumberland and Westmorland Antiquarian and Archaeological Society,* n.s., 60 (1960): 109; H. D. Rawnsley, *A Coach Drive at the Lakes: Windermere to Keswick; and the Buttermere Round* (Keswick: T. Bakewell, 1902), 65.

73. Harriet Martineau, *Guide to Keswick and Its Environs* (Windermere: John Garnett, 1857), 8; Joseph Baron, *All about the English Lakes: A Cyclopedia of Places, Persons, Myths and Happenings* (Kendal: Atkinson and Pollitt, 1925), 14–15.

74. Margaret Armstrong, ed., *Thirlmere across the Bridges to Chapel, 1849–1852: From the Diary of Basil R Lawson, Curate of Wythburn* (Keswick: Peele Wyke, 1989), 6; Mannix and William Whellan, *History, Gazetteer and Directory of Cumberland* (1847; repr., Cumberland: Michael Moon, 1974), 564–65.

75. F. G. Brabant, *The English Lakes* (London: Methuen, 1902), 164; Pevsner, *Cumberland and Westmorland,* 213.

76. Ashley P. Abraham, *Beautiful Lakeland* (Keswick: G. P. Abraham, 1912), 25.

77. [Adam Walker], *A Tour from London to the Lakes: Containing Natural, Oeconomical, and Literary Observations, Made in the Summer of 1791. By a Gentleman* (London: John Abraham, 1792), 107; E. Linn Linton, *The Lake Country* (London: Smith, Elder, 1864), 44; Samuel Barber, *Beneath Helvellyn's Shade: Notes and Sketches in the Valley of Wythburn* (London: Elliot Stock, 1892), 21.

78. James Clarke, *Survey of the Lakes,* 118.

79. Roby X., "Thirlmere, Past and Present, Part I," 31.

80. For detailed surveys of mining at Thirlmere, see generally Alen McFadzean, *Wythburn Mine and the Lead Miners of Helvellyn* (Ulverston, Cumbria: Red Earth Publications, 1987), and Tyler, *Thirlmere Mines.*

81. Harwood, *History and Description,* 38; Thomas Huson, *Round about Helvellyn* (London: Seeley, 1895), 28; Prior, *Guide to the Lake District,* 36; John Robinson, *A Guide*

to the Lakes in Cumberland, Westmorland, and Lancashire (London: Lackington, Hughes, Harding, Mavor, and Jones, 1819), 201–2.

82. Geoffrey Halliday, *A Flora of Cumbria* (Lancaster: Centre for North-West Regional Studies, University of Lancaster, 1997), 45–46; David Hey, "Moorlands," in Joan Thirsk, ed., *The English Rural Landscape* (Oxford: Oxford University Press, 2000), 190; Roy Millward and Adrian Robinson, *The Lake District* (London: Eyre and Spottiswoode, 1970), 72–73.

83. Stuart Piggott, ed., *Prehistory: The Agrarian History of England and Wales,* vol. I-I (Cambridge: Cambridge University Press, 1981), 173–76; Millward and Robinson, *Lake District,* 74–75.

84. W. H. Pearsall and Winifred Pennington, *The Lake District: A Landscape History* (London: Collins, 1973), 231–32.

85. For a detailed listing of prehistoric sites in Cumbria, see Archaeology UK, "The Archaeological Sites Index (ARCHI)" (http://www.digital-documents.co.uk/archi/archi.htm); Geoffrey Clark and W. Harding Thompson, *The Lakeland Landscape* (London: Adam and Charles Black, 1938), 10.

86. R. V. Davis, *Geology of Cumbria: Lakeland's Rocks and Minerals Explained* (Clapham, N. Yorks.: Dalesman Books, 1977), 37; Barry Cunliffe et al., eds., *Penguin Atlas of British and Irish History* (London: Penguin, 2001), 21.

87. Clare Fell, *Early Settlement in the Lake Counties* (Clapham, Yorks.: Dalesman Books, 1972), 65–68; Millward and Robinson, *Lake District,* 122–23; H. C. Darby, "The Anglo-Scandinavian Foundations," in H. C. Darby, ed., *A New Historical Geography of England* (Cambridge: Cambridge University Press: 1973), 13; Fell, *Early Settlement,* 84.

88. Darby, "Anglo-Scandinavian Foundations," 19–20.

89. Fell, *Early Settlement,* 84; Angus J. L. Winchester, *The Harvest of the Hills: Rural Life in Northern England and the Scottish Borders, 1400–1700* (Edinburgh: Edinburgh University Press, 2000), 103–4.

90. Gilpin, *Observations,* x–xii.

91. Smout, *Nature Contested,* 47–48.

92. The relative responsibility of agriculture and manufacturing for deforestation is still matter for disagreement. For example, H. C. Darby characterizes rural industry as a "great devourer of trees," while Oliver Rackham argues forcefully that "the survival of almost any large tract of woodland suggests that there has been an industry to protect it against the claims of farmers." Darby, "The Age of the Improver: 1600–1800," in Darby, *New Historical Geography,* 327; Rackham, *Trees and Woodland in the British Landscape: The Complete History of Britain's Trees, Woods and Hedgerows* (London: Phoenix Press, 1990), 86.

93. W. Hutchinson, *An Excursion to the Lakes in Westmorland and Cumberland* (London: J. Wilkie, 1776), 177; [Budworth/Palmer], *Fortnight's Ramble,* 158; William Whellan, *The History and Topography of the Counties of Cumberland and Westmoreland* (Pontefract: W. Whellan, 1860), 41.

94. W. G. Collingwood, *Lake District History* (Kendal: Wilson and Son, 1925), 123, 128; Alastair Cameron, ed., *Lakeland's Mining Heritage: The Last 500 Years* (Alston, Cumbria: Cumbrian Amenity Trust Mining History Society, 2000), 90–91; J. D. Kendall, "Notes on the History of Mining in Cumberland and North Lancashire," *Transactions of the North of England Institute of Mining and Mechanical Engineers* 34 (1884–85): 116–17; H. D. Rawnsley, *Past and Present at the English Lakes* (Glasgow: James MacLehose, 1916), 67.

95. W. G. Collingwood, *Lake Counties* (New York: E. P. Dutton, 1902), 159.

96. Bouch and Jones, *Short Economic and Social History,* 125.

97. Rawnsley, *Past and Present,* 67, 73.

98. Petition of Stanley Hughes Le Fleming of Rydal Hall, in "Thirlmere. Petitions against Bill. 1878" (Manchester Central Library, M231/2, box 2).

99. House of Commons, Select Committee on the Manchester Corporation Water Bill. Minutes of Evidence (March 14, 1878), 253; Roby X., "Thirlmere, Past and Present. Part II," 52–53.

100. Petition of West Cumberland Consols, Ltd., in "Thirlmere. Petitions against Bill. 1878" (Manchester Central Library, M231/2, box 2).

101. William T. Palmer, *The English Lakes* (London: Adam and Charles Black, 1908), 165.

Chapter 2

1. *Guide to Manchester and Salford* (Edinburgh: Adam and Charles Black, 1882), 3.

2. Daniel Defoe, *A Tour through the Island of Great Britain* (1726; London: J. F. and C. Rivington, 1778), 3:277; James Ogden, *A Description of Manchester by a Native of the Town,* ed. Neil Richardson (Manchester: Neil Richardson, 1983), 6; French observer, quoted in Asa Briggs, *Victorian Cities* (New York: Harper and Row, 1970), 88; Alan Kidd, *Manchester* (Keele: Keele University Press, 1996), 15, 22.

3. Kidd, *Manchester,* 154; for a detailed account of Manchester's geographical aggrandizement, see Arthur Redford, *The History of Local Government in Manchester,* vol. 2, *Borough and City* (London: Longmans, Green, 1940), chap. 24.

4. *An Historical Record of Some Recent Enterprises of the Corporation of Manchester and of Its Co-operation in the Completion of the Manchester Ship Canal* (Manchester: Henry Blacklock, 1894), 9; Kidd, *Manchester,* 105; House of Commons Select Committee on the Manchester Corporation Water Bill (Shorthand Writer's Notes), Minutes of Evidence, testimony of Sir Edmund Beckett QC (March 5, 1878), 6; Martin Daunton, *Progress and Poverty: An Economic and Social History of Britain, 1700–1850* (Oxford: Oxford University Press, 1995), 137.

5. Joseph Aston, *The Manchester Guide: A Brief Historical Description of the Towns of Manchester and Salford, the Public Buildings, and the Charitable and Literary Institutions* (Manchester: Joseph Aston, 1804), 1. Of course this reflected a Mancunian perspective. Residents of Birmingham tended to see things differently.

6. Clare Hartwell, *Manchester* (London: Penguin Books, 2001), 18–19, 89–91, 174–75; John J. Parkinson-Bailey, *Manchester: An Architectural History* (Manchester: Manchester University Press, 2000), 11–13, 63–68. For a detailed account of early Victorian Manchester's literary and scientific societies, see Robert H. Kargon, *Science in Victorian Manchester: Enterprise and Expertise* (Baltimore: Johns Hopkins University Press, 1977), 1–33.

7. H. G. Duffield, *The Strangers Guide to Manchester* (1850; repr., Swinton: Neil Richardson, 1984), 1. For a description of the town hall, see Hartwell, *Manchester,* 71–84; on the architecture of Manchester's Victorian heyday, see Parkinson-Bailey, *Manchester,* chap. 6.

8. Alan Kidd, "The Industrial City and Its Pre-industrial Past: The Manchester Royal Jubilee Exhibition of 1887," *Transactions of the Lancashire and Cheshire Antiquarian Society* 89 (1993): 63–70; *Manchester City News* (June 25, 1887), quoted in Kidd, "Industrial City," 66.

9. For Briggs's extended discussion of Victorian Manchester, see *Victorian Cities,* chap. 3.

10. Thomas Carlyle, *Past and Present* (1843; repr., London: Dent, 1960), 185.

11. James Phillips Kay (Kay-Shuttleworth), *The Moral and Physical Condition of the Working Classes Employed in the Cotton Manufacture in Manchester* (1832; repr., Shannon: Irish University Press, 1971), 46.

12. Kay, *Moral and Physical Condition,* 13, 21, 22, 6.

13. Friedrich Engels, *The Condition of the Working Class in England,* ed. David McLellan, trans. Florence Kelley-Wischnewetsky (1845; Oxford: Oxford University Press, 1993), 54, 58, 63.

14. Parkinson-Bailey, *Manchester,* 1.

15. Aston, *Manchester Guide,* 2.

16. Duffield, *Stranger's Guide,* 1.

17. Johann Georg May, "Report," excerpted in L. D. Bradshaw, ed., *Visitors to Manchester: A Selection of British and Foreign Visitors' Descriptions of Manchester from c1538 to 1865* (Manchester: Neil Richardson, 1987), 25.

18. Alexis de Tocqueville, *Journeys to England and Ireland,* excerpted in Bradshaw, *Visitors to Manchester,* 34.

19. John Frederic LaTrobe Bateman, *History and Description of the Manchester Waterworks* (Manchester: T. J. Day, 1884), 1–2.

20. William Kay, "Presidential Address," British Waterworks Association (17th Annual Meeting, Manchester, July 4–7, 1928), 7.

21. Aston, *Manchester Guide,* 6.

22. Stephen Halliday, *The Great Stink of London: Sir Joseph Bazalgette and the Cleansing of the Victorian Metropolis* (Stroud.: Sutton Publishing, 1999), 32–34; Royal Commission on Water Supply, *Report of the Commissioners* (London: Spottiswoode, 1869), pt. 3, sects. 95–108.

23. F. Hibbert, "A Description of the Liverpool Corporation Water Supply Undertaking—1847–1947," *Water and Water Engineering* 51 (May 1948): 202–3.

24. Michael Taggart, *Private Property and Abuse of Rights in Victorian England: The Story of Edward Pickles and the Bradford Water Supply* (Oxford: Oxford University Press, 2002), 6–7.

25. François Vigier, *Change and Apathy: Liverpool and Manchester during the Industrial Revolution* (Cambridge, MA: MIT Press, 1970), 125; Shena D. Simon, *A Century of City Government: Manchester 1838–1938* (London: George Allen and Unwin, 1938), 349.

26. Report quoted in Bateman, *History and Description,* 5.

27. Vigier, *Change and Apathy,* 125.

28. Simon, *Century of City Government,* 350.

29. Bateman, *History and Description,* 9.

30. Ibid., 12–18; Vigier, *Change and Apathy,* 195; William E. A. Axon, *The Annals of Manchester: A Chronological Record from the Earliest Times to the End of 1885* (Manchester: John Heywood, 1886), 144, 149, 153, 162, 167, 213.

31. Harold L. Platt, *Shock Cities: The Environmental Transformation and Reform of Manchester and Chicago* (Chicago: University of Chicago Press, 2005), 204–5.

32. J. A. Hassan, "The Growth and Impact of the British Water Industry in the Nineteenth Century," *Economic History Review* 38 (1985): 531–47, esp. 544–45; William Ralph Baldwin-Wiseman, "The Increase in the National Consumption of Water," *Journal of the Royal Statistical Society* 72 (1909): 251.

33. Simon, *Century of City Government,* 349; City of Manchester, *Centenary of the Waterworks Undertaking 1847–1947* (Manchester, 1947), 32.

34. London's water problems have received a great deal of historical attention. See, for example, Halliday, *Great Stink;* Dale H. Porter, *The Thames Embankment: Environment, Technology and Society in Victorian London* (Akron: University of Akron Press, 1998); Christopher Hamlin, *A Science of Impurity: Water Analysis in Nineteenth-Century Britain* (Berkeley: University of California Press, 1990); and Bill Luckin, *Pollution and Control: A Social History of the Thames in the Nineteenth Century* (Bristol: Adam Hilger, 1986).

35. Hassan, "Growth and Impact," 533–34.

36. Simon, *Century of City Government,* 20.

37. Ibid., 67–69; Kidd, *Manchester,* 64; Redford, *History of Local Government,* 2:6–9.

38. "The Mosley Family of Manchester," http://www.thornber.net/cheshire/htmlfiles/mosley.html.

39. See Derek Fraser, "Introduction: Municipal Reform in Historical Perspective," in Derek Fraser, ed. *Municipal Reform and the Industrial City* (Leicester: Leicester University Press, 1982), 3–4.

40. Redford, *History of Local Government,* 2:9–26. On the political divisions during this period, see V. A. C. Gatrell, "Incorporation and the Pursuit of Liberal Hegemony in Manchester 1790–1839," in Fraser, *Municipal Reform,* 16–60.

41. Redford, *History of Local Government,* 2:28; Simon, *Century of City Government,* 395.

42. "The Late Sir Joseph Heron, First Town Clerk of Manchester," *Manchester Faces and Places* 1 (January 10, 1890), 49; Simon, *Century of City Government,* 407.

43. "The Late Sir Joseph Heron," 51–52.

44. Bateman, *History and Description,* 19.

45. Ibid., 23.

46. John James Harwood, *A Short Description of the Manchester Corporation Waterworks* (Manchester: H. Blacklock, 1891), 6. For extended accounts of the struggles of the Manchester Corporation with the Etherow valley mill owners, see Redford, *History of Local Government,* vol. 2, chap. 20, and, from a different perspective, Tom Quayle, *The Cotton Industry in Longdendale and Glossopdale* (Stroud: Tempus, 2006), chap. 10.

47. Bateman, *History and Description,* 56–57; Redford, *History of Local Government,* 2:177–82.

48. Harwood, *Short Description,* 12–13; James Mansergh, *The Thirlmere Scheme of the Manchester Corporation, with a Few Remarks on the Longdendale Works and Water Supply Generally* (London: E and F. N. Spon, 1878), 12.

49. Bateman, *History and Description,* 90.

50. *Proceedings of the Manchester Borough Council,* 22nd July 1846, quoted in Redford, *History of Local Government,* 2:177.

51. Laurence E. Breeze, *The British Experience with River Pollution, 1865–1876* (New York: Peter Lang, 1993), 98.

52. Bateman, *History and Description,* 86–87.

53. Quoted in ibid., 95.

54. John C. Thresh, *Water and Water Supplies* (1896; Philadelphia: P. Blakiston's Son, 1901), v, 307.

55. "The Thirlmere Waterworks," *Health Journal* (February 1886), 137; Thresh, *Water and Water Supplies,* 313.

56. Thresh, *Water and Water Supplies,* 307.

57. Simon, *Century of City Government,* 288.

58. Alan Wilson, "Technology and Municipal Decision-Making: Sanitary Systems in Manchester 1868–1910" (PhD diss., Manchester University, 1990), 69, 170. For a trenchant critique, see Platt, *Shock Cities,* esp. 70–71.

59. Marilyn E. Pooley and Colin G. Pooley, "Health, Society and Environment in Victorian Manchester," in *Urban Disease and Mortality in Nineteenth-Century England,* ed. Robert Woods and John Woodward (London: Batsford, 1984), 175.

60. Duffield, *Stranger's Guide,* 16; Redford, *History of Local Government,* 2:176.

61. *Guide to Manchester and Salford,* 22.

62. *Historical Record,* 115–16.

63. House of Commons Select Committee on the Manchester Corporation Water Bill (Shorthand Writer's Notes), Minutes of Evidence (March 5, 1878), Sir Joseph Heron, 15.

64. Ibid., Minutes of Evidence (March 5, 1878), Sir Edmund Beckett, 9.

65. *Manchester Guardian* (February 10, 1852), quoted in Tom Quayle, *Reservoirs in the Hills: The Story of the Construction of the Reservoirs in the Longdendale Valley Which Feed Manchester with Water* (n.p.: Senior Publications, 1988), 30; Peter Russell, "Bateman, John Frederic LaTrobe (1810–1889)," *Oxford Dictionary of National Biography* (Oxford: Oxford University Press, 2004).

66. Bateman, *History and Description,* preface.

67. "The Water Supply of Manchester. Completion of the Works in Longdendale," *Manchester Guardian* (January 31, 1877); Minutes of the Waterworks Committee, 1878, quoted in J. A. Hassan and E. R. Wilson, "The Longdendale Water Scheme 1848–1884," *Industrial Archaeology* 14 (1979): 103.

68. Bateman, *History and Description,* preface.

69. *A Microscopical Examination of Certain Waters Submitted to Jabez Hogg, Surgeon to the Royal Westminster Ophthalmic Hospital . . . and a Chemical Analysis by Dugald Campbell, Analytical Chemist to the Brompton Hospital* (London: Balliere, Tindall, Cox, 1874), 19.

70. Bateman's report to the Manchester and Salford Waterworks Company, 1844, quoted in Bateman, *History and Description,* 42; Harwood, *Short Description,* 4.

71. Hassan and Wilson, "Longdendale Water Scheme," 104, 106.

72. James Winter, *Secure from Rash Assault: Sustaining the Victorian Environment* (Berkeley: University of California Press, 1999), 171–72; *Manchester Guardian* (February 10, 1852), quoted in Quayle, *Reservoirs in the Hills,* 30.

73. Quayle, *Reservoirs in the Hills,* 22; Hassan and Wilson, "Longdendale Water Scheme," 106–7.

74. Bateman, quoted in Quayle, *Reservoirs in the Hills,* 37.

75. Redford, *History of Local Government,* 2:202.

76. Waterworks subcommittee, quoted in Quayle, *Reservoirs in the Hills,* 43; ibid., 36.

77. According to Hassan and Wilson, such onerous and, in their view, overgenerous arrangements were frequent in the early years of municipal control of waterworks. Hassan and Wilson, "Longdendale Water Scheme," 111; J. H. Balfour Browne, *Water Supply* (London: Macmillan, 1880), 14.

78. Redford, *History of Local Government,* 2:202; John James Harwood, *History and Description of the Thirlmere Water Scheme* (Manchester: Henry Blacklock, 1895), 5; Bateman, *History and Description,* 48, 50.

79. Mansergh, *Thirlmere Scheme,* 20.

80. Bateman, *History and Description,* 151.

81. Hassan and Wilson, "Longdendale Water Scheme," 117.

82. *Guide to Manchester and Salford,* 37; Quayle, *Reservoirs in the Hills,* 46–47.

83. Hassan and Wilson, "Longdendale Water Scheme," 116.

84. Harwood, *History and Description,* 6.

85. "Water Supply of Manchester," *Manchester Guardian* (January 31, 1877); Mansergh, *Thirlmere Scheme,* 20.

86. Bateman, quoted in Harwood, *History and Description,* 7; letter from Bateman to Heron, May 24, 1875, appendix 2 in ibid., 199–200.

87. Browne, *Water Supply* , 15.

Chapter 3

1. Royal Commission on Water Supply, "Original Commission," in *Report of the Commissioners* (London: Spottiswoode, 1869), n.p.

2. Royal Commission on Water Supply, *Report,* pt. 1, xix.

3. Ibid., xxii.

4. M.C.W.W. Supply of Water and Proposed New Works. Sub-Committee Minute Book (October 6, 1874), 1 (Manchester Central Library, M231/2/2, box 4).

5. Ibid. (March 11, 1875), 1.

6. John Frederic LaTrobe Bateman, *History and Description of the Manchester Waterworks* (Manchester: T. J. Day, 1884), 206.

7. "The Thirlmere Water Scheme," *Manchester Courier* (October 31, 1877), reprinted in Manchester Corporation Waterworks, *Supply of Water from Thirlmere. Extracts from the Manchester Newspapers* (Manchester: A. Ireland, 1877), 4.

8. Bateman, *History and Description,* 214. For a brief overview of Liverpool's water history, see F. Hibbert, "A Description of the Liverpool Corporation Water Supply Undertaking—1847–1947," *Water and Water Engineering* (May 1948): 202–11.

9. Grave did not receive enduring credit for his role in the Thirlmere Scheme. In 1920, a pseudonymous admirer labeled him "A Forgotten Citizen," before attempting to rescue him from obscurity: "And yet to thee we owe the stream/Of purest water that doth come/From Thirlmere to our city home." Carlos, "A Forgotten Citizen," *Manchester City News,* November 13, 1920 (in scrapbook, "M.C.W.W. Newspaper Cuttings," Manchester Central Library, M231/125).

10. M.C.W.W. Supply of Water and Proposed New Works. Sub-Committee Minute Book (October 12, 1877), 125.

11. "The Thirlmere Water Scheme," *Manchester Examiner and Times* (October 31, 1877), reprinted in Manchester Corporation Waterworks, *Supply of Water,* 4.

12. J. Wilson, *Thirlmere to Manchester: A Compendious History of the Promotion, Progress, and Construction of the Great Thirlmere Waterworks Scheme* (Ambleside: George Middleton, 1894), 4.

13. "The Thirlmere Water Scheme," *Manchester Courier* (October 31, 1877), reprinted in Manchester Corporation Waterworks, *Supply of Water,* 5.

14. Roby X., "Thirlmere, Past and Present. Part III," *North Lonsdale Magazine and Furness Miscellany* 4 (December 1900), 76.

15. M.C.W.W. Supply of Water and Proposed New Works. Sub-Committee Minute Book (July 13 and August 2, 1876), 3, 6.

16. John James Harwood, *History and Description of the Thirlmere Water Scheme* (Manchester: Henry Blacklock, 1895), 27.

17. *26th, 27th, and 28th September, 1894. Manchester Corporation Waterworks Act 1879. Thirlmere Aqueduct. Lord Lonsdale's Estates. Arbitration between the Right Hon. James Lowther, M.P. and the Corporation of Manchester. Before Christopher Oakley, Esq. Umpire Appointed by the Board of Trade* (Manchester: Henry Blacklock, 1894), 74.

18. Harwood, *History and Description,* 36.

19. Documents relating to *Vane v. Vane* (Cumbria Record Office, Carlisle, S/NWA/1/22).

20. M.C.W.W. Supply of Water and Proposed New Works. Sub-Committee Minute Book, 7, 17–18.

21. "Manchester Corporation Waterworks. Thirlmere. Purchase from Sir Henry Vane of the Manor of Wythburn. Memoranda of Interviews, Correspondence, etc." (Cumbria Record Office, Carlisle, S/NWA/3/11), 2, 3.

22. Harwood, *History and Description,* 59; letter from Sir Henry Vane's solicitor to John Grave (February 2, 1878) (Cumbria Record Office, Carlisle, S/NWA/18).

23. C. Roy Huddleston and R. S. Boumphrey, *Cumberland Families and Heraldry* (Cumberland and Westmorland Antiquarian and Archaeological Society, 1978), 202; Bruce L. Thompson, "The Leathes Family of Dalehead," *Transactions of the Cumberland and Westmorland Antiquarian and Archaeological Society,* n.s., 60 (1960): 109; Roby X., "Thirlmere, Past and Present. Part I," *North Lonsdale Magazine and Furness Miscellany* (August 1900), 31.

24. Harwood, *History and Description,* 56–57.

25. M.C.W.W. Supply of Water and Proposed New Works. Sub-Committee Minute Book, 53–54.

26. House of Commons Select Committee on the Manchester Corporation Water Bill (Shorthand Writer's Notes), Minutes of Evidence (March 19, 1878), 327; M.C.W.W. Supply of Water and Proposed New Works. Sub-Committee Minute Book, 65.

27. Letters to the *Standard* (November 6, November 8, and November 9, 1877); "Manchester v. Thirlmere," *Times* (December 18, 1877), 6.

28. Petition of Leonard Stanger Leathes, 2, in "Thirlmere. Petitions against Bill. 1878" (Manchester Central Library, M231/2, box 2).

29. House of Commons Select Committee on the Manchester Corporation Water Bill (Shorthand Writer's Notes), Minutes of Evidence (March 19, 1878), 330.

30. Ibid., Opening Statement (March 5, 1878), 13.

31. "The Tapping of Thirlmere," *New Quarterly Magazine* (April 1878), 207.

32. House of Commons Select Committee on the Manchester Corporation Water Bill (Shorthand Writer's Notes), Opening Statement (March 5, 1878), 13.

33. Ian Tyler, *Thirlmere Mines and the Drowning of the Valley* (Keswick: Blue Rock Publications, 1999), 153–54.

34. M.C.W.W. Supply of Water and Proposed New Works. Sub-Committee Minute Book (November 20 and December 6, 1876), 21.

35. Ibid. (December 8, 1876), 25; Harwood, *History and Description,* 51–52, 214; Thomas Barclay, *The Future Water Supply of Birmingham* (Birmingham: Cornish Brothers, 1898), 200–205.

36. Harwood, *History and Description,* 59.

37. "Manchester Corporation Waterworks. Thirlmere. Purchase from Sir Henry Vane of the Manor of Wythburn. Memoranda of Interviews, Correspondence, etc.," 1–2; Harwood, *History and Description,* 50.

38. Douglas Sutherland, *The Yellow Earl: The Life of Hugh Lowther, 5th Earl of Lonsdale, 1857–1944* (1965; Exeter: Molendinar Press, 1980), 63–65, 107.

39. Ibid., 276; *Manchester Corporation Waterworks Act 1879. Thirlmere Aqueduct. Lord Lonsdale's Estates. Arbitrations between the Right Hon. James Lowther, M. P. And the Corporation of Manchester. Awards of Christopher Oakley, Esq. (Umpire Appointed by the Board of Trade). December 21st, 1894* (Manchester: Henry Blacklock, 1894), 7, 11.

40. House of Commons Select Committee on the Manchester Corporation Water Bill (Shorthand Writer's Notes), Minutes of Evidence (March 20, 1878), 368–69, 464; Manchester City Council, *The Thirlmere Water Supply Scheme. Speeches of Mr. Alderman King and Mr. Alderman Curtis* (London: C. F. Roworth, 1878), 15.

41. "August 16th 1878. Thirlmere Water Bill. Report of Meeting of Owners and Ratepayers in the Town Hall," 3–6 (Manchester Central Library, M231, box 2).

42. Ibid., 49.

43. Ibid., 75, 79.

44. "Cost of Poll in Thirlmere Scheme," Manchester Election Expenses, 63 (Manchester Central Library, MSC 268); "Poll taken pursuant to public meeting of 16 August 1878," Statistical Returns. 1879 (Manchester Central Library, M231/2, box 8).

45. John King, *Manchester and Its Waterworks. Where Are We? A Letter on the Present Position of the Question, and Especially on Its Financial Aspect, to His Fellow-Members of the Manchester City Council* (Manchester: Charles Sever, 1884), 3.

46. January 4, 1878, "Thirlmere Defence Association Account Book," 9 (Manchester Central Library, M48, box 9).

47. "August 16th 1878. Thirlmere Water Bill. Report of Meeting of Owners and Ratepayers in the Town Hall," 1–2 (Manchester Central Library, M231, box 2).

48. John S. Clarke, *The Epic of Municipalisation: The Story of Glasgow's Loch Katrine Water Supply* (Glasgow: Forward, 1928), cover, 21–26.

49. James M. Gale, *Papers Read before the Institution of Engineers in Scotland on the Glasgow Waterworks* (Glasgow: Gardner and Stevenson, 1864), 10–11. For general responses to the Loch Katrine project, see T. C. Smout, *Nature Contested: Environmental History in Scotland and Northern England since 1600* (Edinburgh: Edinburgh University Press, 2000), 111–12, and Irene Maver, "Children and the Quest for Purity in the Nineteenth Century Scottish City," *Paedagogica Historica* 33 (1997): 810–11.

50. Robert Somervell, *Chapters of Autobiography* (London: Faber and Faber, 1935), 51; Tim Hilton, *John Ruskin: The Later Years* (New Haven: Yale University Press, 2000), 335; John Ruskin, preface to Robert Somervell, *A Protest against the Extension of Railways in the Lake District* (Windermere: J. Garnett, 1876), 1.

51. Somervell, *Protest against the Extension*, 21, 26–27.

52. Somervell, *Chapters of Autobiography*, 51.

53. Tyler, *Thirlmere Mines*, 146.

54. Thirlmere Defence Association (November 23, 1877) (Cumbria Record Office, Carlisle, Bibliotheca Jacksoniana).

55. *Water for Manchester from Thirlmere. The Manchester and Thirlmere Scheme: An Appeal to the Public on the Facts of the Case* (Windermere: J. Garnett, 1877), 2.

56. Somervell, *Chapters of Autobiography*, 52.

57. *Water for Manchester from Thirlmere*, 4.

58. Ibid., 9.

59. Ibid., 10.

60. Thirlmere Defence Association, *Is Thirlmere the Only Source of Supply for Manchester* (London: Sir Joseph Causton, 1878), 7.

61. *Thirlmere Defence Association. The Case Re-stated* (Windermere: J. Garnett, 1878), 13.

62. Ibid., 15–16.

63. "Manchester and Thirlmere," *Times* (December 22, 1877), 3; Somervell, *Chapters of Autobiography,* 53.

64. James Clifton Ward, *The Geology of the Northern Part of the English Lake District* (London: HMSO, 1876).

65. Harvey Carlisle, "Manchester v. Thirlmere," *Times* (October 20, 1877), 11; Robert T. Farquhar, "Manchester v. Thirlmere," *Times* (October 31, 1877) 6; John Grave, "Manchester and Thirlmere," *Times* (December 15, 1877), 4; Octavia Hill, "To the Editor of the *Times,*" *Times* (December 21, 1877), 5.

66. *Times* (October 20, 1877), *Daily News* (November 2, 1877), *Standard* (November 2, 1877), and *Pall Mall Gazette* (November 8, 1877), all reprinted in Thirlmere Defence Association, *Extracts from the Leading Journals on the Manchester Water Scheme* (Windermere: J. Garnett, 1878), 5, 6, 7, 9.

67. M.C.W.W. Supply of Water and Proposed New Works. Sub-Committee Minute Book (May 3, 1877), 60–61.

68. *Yorkshire Post and Leeds Intelligencer* (January 2, 1878), reprinted in TDA, *Extracts,* 15.

69. "Manchester and the Meres," *Spectator* (September 8, 1877), 118.

70. *Field* (November 17, 1877), *Medical Press and Circular* (January 2, 1878), and *Builder* (December 1, 1877), all reprinted in TDA, *Extracts,* 14, 16, 14.

71. John Ruskin, *Fors Clavigera: Letters to the Workmen and Labourers of Great Britain* (1871–84; repr., Boston: Colonial Press, 1900) , 2:74 (letter 79, June 18, 1877), 2:138–39 (letter 82, September 13, 1877).

72. "A Talk by Thirlmere," *Punch* 74 (February 9, 1878), 49; "Boring for Water," *Punch* (August 31, 1878), 96.

73. John Wilson Grisdale, *"A Voice from the Ranks": An Original Poem on Thirlmere Lake* (Keswick: Bakewell and Fitzjames, 1877), 4.

74. An Oddfellow [pseud.], *King Croesus; or, Harlequin Town Clerk Irwell and the Thirlmere Lake (An Unacted Fairy Extravaganza)* (London: Samuel French, 1878), 22.

75. Letter from Joseph Heron to the town clerk of Carlisle (January 9, 1878) (Manchester Central Library, M48, box 11).

76. Ben Brierley, *A Trip to Thirlmere and Borrowdale* (Manchester: Abel Heywood, 1878), 4, 6, 11.

77. "The Prince's Pantomime," *City Jackdaw* (December 21, 1877), 48.

78. "Mr Ben Brierley," *Manchester Faces and Places* 1 (September 10, 1890): 177–80; "Benjamin Brierley," *Dictionary of National Biography.*

79. "The Thirlmere Water Scheme," *Manchester Examiner and Times* (October 31, 1877), "The Thirlmere Water Scheme," *Manchester Courier* (October 31, 1877), and "The Thirlmere Water Scheme," *Manchester Guardian* (October 31, 1877), all reprinted in Manchester Corporation Waterworks, *Supply of Water,* 7, 9, 11, 1.

80. *Daily News* (October 25, 1877), reprinted in Harwood, *History and Description,* 233.

81. "Grundy on the Box," *City Jackdaw* (November 9, 1877), between pp. 411 and 412.

82. Michael Taggart, *Private Property and Abuse of Rights in Victorian England: The Story of Edward Pickles and the Bradford Water Supply* (Oxford: Oxford University Press, 2002), 18; *An Historical Record of Some Recent Enterprises of the Corporation of Manchester*

and of Its Co-operation in the Completion of the Manchester Ship Canal (Manchester: Henry Blacklock, 1894), 55.

83. Frederick Clifford, *A History of Private Bill Legislation* (London: Butterworths, 1885), 1:267–69.

84. *London Gazette,* no. 24525 (November 23, 1877), 6569–71.

85. Clifford, *History of Private Bill Legislation,* 2:480.

86. House of Commons Select Committee on the Manchester Corporation Water Bill (Shorthand Writer's Notes), Opening Statement (March 5, 1878), 1.

87. *Hansard Parliamentary Debates,* 3rd ser., vol. 237 (February 12, 1878), cols. 1503–4.

88. Ibid., cols. 1508, 1512, 1527.

89. Ibid., cols. 1517, 1524.

90. *Lloyd's Weekly Newspaper* (February 17, 1878).

91. *Hansard* 237 (February 15, 1878), cols. 1524, 1723.

92. Letter from Messrs. Harward and Co. to the town clerk, Manchester (October 22, 1877) (Manchester Central Library, M48, box 11).

93. Somervell, *Chapters of Autobiography,* 52.

94. Ibid., 55.

95. House of Commons Select Committee on the Manchester Corporation Water Bill (Shorthand Writer's Notes), Minutes of Evidence (March 15, 1878), 287.

96. Harwards, Shepherd and Mills, "Manchester Corporation Water Bill. Instructions for Cross-Examination of Mr. Grave," 1878 (Manchester Central Library, M48, box 11).

97. Harwards, Shepherd and Mills, "Manchester Corporation Water Bill. Notes for Cross Exam of Sir Jos. Heron," 1878 (Manchester Central Library, M48, box 11).

98. Harwards, Shepherd and Mills, "Manchester Corporation Water Bill. Instructions for Cross Examination of Mr. Bateman," 1878 (Manchester Central Library, M48, box 11).

99. House of Commons Select Committee on the Manchester Corporation Water Bill (Shorthand Writer's Notes), Minutes of Evidence (March 12, 1878), 145–46.

100. Ibid., Minutes of Evidence (March 13, 1878), 236.

101. Ibid., Minutes of Evidence (March 12, 1878), 159.

102. Ibid., Minutes of Evidence (March 13, 1878) , 204–5, 212.

103. Ibid., Minutes of Evidence (March 12, 1878), 156.

104. Ibid., Minutes of Evidence (March 13, 1878), 224.

105. Ibid., Opening Remarks (March 5, 1878), 18.

106. Royal Commission on Water Supply, *Report,* pt. 1, xxii.

107. House of Commons Select Committee on the Manchester Corporation Water Bill (Shorthand Writer's Notes), Speeches (March 27, 1878), 16.

108. Ibid., Minutes of Evidence (March 27, 1878), 510.

109. M.C.W.W. Supply of Water and Proposed New Works. Sub-Committee Minute Book (November 22, 1877), 131 (Manchester Central Library, M231/2/2, box 4).

110. Thirlmere Defence Association, *Case Re-stated,* 16.

111. House of Commons Select Committee on the Manchester Corporation Water Bill (Shorthand Writer's Notes), Minutes of Evidence (March 14, 1878), 272, 274.

112. Ibid., Mr. Pember's Speech (March 19, 1878), 1.

113. Ibid., Mr. Pember's Speech (March 19, 1878), 5, 10.

114. Ibid., Minutes of Evidence (March 15, 1878), 288, 286.

115. Ibid., Minutes of Evidence (March 15, 1878), 292, 288.

116. Ibid., Minutes of Evidence (March 15, 1878), 293.

117. Ibid., Minutes of Evidence (March 15, 1878), 291.

118. Ibid., Minutes of Evidence (March 15, 1878), 297.

119. Ibid., Minutes of Evidence (March 19, 1878), 333.

120. Ibid., Minutes of Evidence (March 22, 1878), 458.

121. Ibid., Minutes of Evidence (March 13, 1878), 215.

122. Ibid., Minutes of Evidence (March 22, 1878), 460; ibid., Sir Edmund Beckett's Speech (March 28, 1878), 3–4.

123. *Manchester and Thirlmere Water Scheme: The Case of the Thirlmere Defence Association* (Windermere: J. Garnett, 1878), 9.

124. Oddfellow, *King Croesus,* 5.

125. *Report from the Select Committee on the Manchester Corporation Water Bill; with the Proceedings of the Committee,* April 8, 1878, iv–v.

126. Credat Judaeus [pseud.], *Thirlmere Water Scheme. Remarks on the Evidence Given before the Select Committee of the House of Commons* (London: P. S. King, 1878), 3.

127. Ibid., 3–4, 21.

128. Ibid., 14, 20, 17.

129. Thirlmere Defence Association, *Report to the Subscribers to the above Association upon the Manchester Corporation Water Works Bill for obtaining a Supply of Water from Lake Thirlmere* (1878), 3–4.

130. Harwood, *History and Description,* 106.

131. Somervell, *Chapters of Autobiography,* 56; John Ruskin, *Works,* ed. E. T. Cook and Alexander Wedderburn (CD ROM version; Cambridge: Cambridge University Press, 1996), 13:516–17.

132. Printed "Report of the Waterworks Committee to the Manchester City Council" (August 8, 1878), inserted into M.C.W.W. Supply of Water and Proposed New Works. Sub-Committee Minute Book (Manchester Central Library, M231/2/2, box 4).

133. "The Thirlmere Water Works: Interview with the Engineer," *Manchester Weekly Times* (October 5, 1894), 22.

134. Manchester Corporation Waterworks Act. 42 & 43 Vict. c.xxxvi (May 23, 1879).

135. "Manchester Corporation Water. Petition of the Thirlmere Defence Association. Against, by Counsel. House of Commons." 1879 (Manchester Central Library, M48, box 1).

136. Manchester Corporation Waterworks Bill. 42 & 43 Vict.—Sess. 1878–79 (copy with longhand annotations, Manchester Central Library, M48, box 4).

137. M.C.W.W. Thirlmere Scheme. Directions Given by Sub-Committee (October 31, 1878), 18 (Manchester Central Library, M231/Special S. C. Thirlmere, box 2).

138. M.C.W.W. Supply of Water and Proposed New Works. Sub-Committee Minute Book (March 17 and April 5, 1879), 200 (Manchester Central Library, M231/2/2, box 4).

139. Wilson, *Thirlmere to Manchester,* 37.

140. *Thirlmere Defence Association* (August 1879).

141. Somervell, *Chapters of Autobiography,* 56.

Chapter 4

1. John James Harwood, *History and Description of the Thirlmere Water Scheme* (Manchester: Henry Blacklock, 1895), 137; *An Historical Record of Some Recent Enterprises of the Corporation of Manchester and of Its Co-operation in the Completion of the Manchester Ship Canal* (Manchester: Henry Blacklock, 1894), 56.

2. John King, *Manchester and Its Waterworks. Where Are We? A Letter on the Present Position of the Question, and Especially on Its Financial Aspect, to His Fellow-Members of the Manchester City Council* (Manchester: Charles Sever, 1884), insert between pp. 16 and 17. King's statistics were drawn from G. J. Symons, *British Rainfall* (1882).

3. King, *Manchester and Its Waterworks,* 18.

4. J. Wilson, *Thirlmere to Manchester: A Compendious History of the Promotion, Progress, and Construction of the Great Thirlmere Waterworks Scheme* (Ambleside: George Middleton, 1894), 37.

5. Harold L. Platt, *Shock Cities: The Environmental Transformation and Reform of Manchester and Chicago* (Chicago: University of Chicago Press, 2005), 393–95.

6. Harwood, *History and Description,* 138–40.

7. "Mr. G. H. Hill, M. Inst. C. E.," *Manchester Faces and Places* 6 (December 1894), 43–44.

8. "Waterworks Committee Report for 1885," 6; "Waterworks Committee Report for 1886," 4–5 (both in Greater Manchester County Record Office, M.C.W.W. Annual Statements).

9. "Waterworks Committee Report for 1887," 4 (Greater Manchester County Record Office, M.C.W.W. Annual Statements).

10. Wilson, *Thirlmere to Manchester,* 39, 40.

11. Ian Tyler, *Thirlmere Mines and the Drowning of the Valley* (Keswick: Blue Rock Publications, 1999), 179; Wilson, *Thirlmere to Manchester,* 46–48.

12. Harwood, *History and Description,* 147.

13. Joseph Scott, *Leaves from the Diary of a Citizen Auditor* (Manchester: Office of the Manchester City News, 1894), 42–50.

14. Wilson, *Thirlmere to Manchester,* 49–50.

15. Harwood, *History and Description,* 147. Harwood described this presentation in the third person, with no indication that the writer was the recipient of the accolades.

16. Wilson, *Thirlmere to Manchester,* 50.

17. Platt, *Shock Cities,* 395.

18. Tyler, *Thirlmere Mines,* 175.

19. Ibid., 158–59.

20. Reginald Blunt, "The Translation of Thirlmere," *Pall Mall Magazine* 3 (May 1894), 19.

21. Harwood, *History and Description,* 164–67.

22. House of Commons Select Committee on the Manchester Corporation Water Bill (Shorthand Writer's Notes), Mr. Maddison's Speech (March 27, 1878), 42; M.C.W.W. Thirlmere Scheme, Directions Given by Sub-Committee (March 6, 1879), 63 (Manchester Central Library, M231/Special S. C. Thirlmere, box 2).

23. Harwood, *History and Description,* 146. For example, the Cumberland County Council reported that the Manchester Corporation had authorized the appointment of an additional constable in the Kirkby Lonsdale district. Cumberland County Council, Minutes of Council and Committees, 1:144 (August 7, 1889) (Cumbria Record Office, Carlisle).

24. "The Thirlmere Waterworks. Serious Quarrel Among Navvies," *Leeds Mercury* (September 17, 1890).

25. H. D. Rawnsley, "To the Workmen," in *Commemorative Sonnets. To the Chairman and the Members of the Waterworks Committee of the Manchester Corporation, the Engineers, Builders, and Workmen, on the Occasion of the Opening of the Thirlmere Waterworks, Friday,*

12th October, 1894 (n.p., 1894). On working conditions and the deaths and injuries that occurred in the course of construction, see Tyler, *Thirlmere Mines,* 174–76, 182–84.

26. Roby X., "Thirlmere Past and Present. Part III," *North Lonsdale Magazine and Furness Miscellany* 4 (December 1900), 75; Samuel Barber, *Beneath Helvellyn's Shade: Notes and Sketches in the Valley of Wythburn* (London: Elliot Stock, 1892), 5–6.

27. Letter of John Dunne, chief constable of Cumberland and Westmorland, quoted in Harwood, *History and Description,* 147; *Westmorland Gazette* (October 13, 1894), n.p.

28. Tim Capelli, *The Thirlmere Way: A Long Distance Walk from Manchester to the Lake District* (Wilmslow, Cheshire: Sigma Press, 1992), 111.

29. M. J. B. Baddeley, *The English Lake District,* 6th ed., Thorough Guide Series (London: Dulau, 1891), 57.

30. Blunt, "Translation of Thirlmere," 7–8.

31. Ibid., 7, 8, 9, 12, 18.

32. Ibid., 7.

33. Roby X., "Thirlmere Past and Present. Part III," 74, 78.

34. Ibid., 78–79.

35. Alen McFadzean, *Wythburn Mine and the Lead Miners of Helvellyn* (Ulverston, Cumbria: Red Earth Publications, 1987), 45–47; Tyler, *Thirlmere Mines,* 109–12.

36. Tyler, *Thirlmere Mines,* 215.

37. Barber, *Beneath Helvellyn's Shade,* 21, 23; Blunt, "The Translation of Thirlmere," 13.

38. H. D. Rawnsley, *Past and Present at the English Lakes* (Glasgow: James Maclehose, 1916), 209–10.

39. "The Climbs of the English Lake District. Part II," *All the Year Round* (November 8, 1884), 114.

40. H. D. Rawnsley, "The Proposed Permanent Lake District Defense Society," *Transactions of the Cumberland Association for the Advancement of Literature and Science* 7 (1882–83): 77–80.

41. Thomas V. Welch, *How Niagara Was Made Free: The Passage of the Niagara Reservations Act in 1885* (Buffalo: Niagara Frontier Historical Society, 1902), 25.

42. "The Lake District Defense Society," Ambleside, January 1885 (Armitt Library, LD1/56).

43. J. D. Marshall and John K. Walton, *The Lake Counties from 1830 to the Mid-Twentieth Century: A Study in Regional Change* (Manchester: Manchester University Press, 1981), 214–15.

44. Robert Farquhar, *Objections to the Thirlmere Scheme (Manchester Water Bill. 1879)* (Ambleside: "Lakes Chronicle" Office, 1879), 12; House of Commons Select Committee on the Manchester Corporation Water Bill (Shorthand Writer's Notes), Minutes of Evidence (March 12, 1878), 156.

45. T. T. Macan, "Life in the Water," in G. A. K. Hervey and J. A. G. Barnes, *Natural History of the Lake District* (London: Frederick Warne, 1970), 82–83.

46. Joseph Nicolson and Richard Burn, *The History and Antiquities of the Counties of Westmorland and Cumberland* (London: W. Strahan and T. Cadell, 1777), 2:79; James Clarke, *A Survey of the Lakes of Cumberland, Westmorland, and Lancashire* (London: James Clarke, 1787), 117.

47. Henry Irwin Jenkinson, *Jenkinson's Practical Guide to the English Lake District* (London: Edward Stanaford, 1879), 59; John Davy, *The Angler in the Lake District; or, Piscatory Colloquies and Fishing Excursions in Westmoreland and Cumberland* (London: Longman, Brown, Green, Longmans and Roberts, 1857), 214.

48. House of Commons Select Committee on the Manchester Corporation Water Bill (Shorthand Writer's Notes), Minutes of Evidence (March 6, 1878), 33.

49. M.C.W.W. Thirlmere Scheme. Directions Given by Sub-Committee (September 19, 1878), 1 (Manchester Central Library, M231/Special S. C. Thirlmere, box 2).

50. Ibid. (April 18, 1879), 76; *Eighteenth Annual Report of the Inspectors of Salmon Fisheries (England and Wales). (For the Year 1878)* (London: HMSO, 1879), 27.

51. George Foster Braithwaite, *The Salmonidae of Westmorland, Angling Reminiscences, and Leaves from an Angler's Notebook* (Kendal: Atkinson and Pollitt, 1884), 47, 56; John Watson, *The English Lake District Fisheries* (London: Lawrence and Bullen, 1899), 129–39; *The English Lake District,* Concise Series of Guides, no. 2 (Kendal: Titus Wilson, 1908), 78; Ernest Philips, "A Guide to Fishing for Trout in British Lakes and Reservoirs," *Fishing Gazette* (December 2, 1922).

52. Barber, *Beneath Helvellyn's Shade,* 77; Watson, *English Lake District Fisheries,* 61.

53. *Ferguson's Tourists' Guide to Callender, the Trosachs, Loch Katrine, Loch Lomond . . .* (Glasgow: Thomas Murray, 1911), 34; John Baynes and George Westropp, *Lake Vyrnwy: The Story of a Sporting Hotel* (Oswestry: Kettering Books, 1992), 51; Walter M. Gallichan [Geoffrey Mortimer, pseud.], *Fishing in Wales: A Guide to the Angler* (London: F. E. Robinson, 1903), 136.

54. Harwards, Shepherd and Mills, "Manchester Corporation Water Bill. Instructions for Cross Examination of Mr. Bateman," 1878, 6 (Manchester Central Library, M48, box 11).

55. Berry to T. McKenny Hughes (February 23, 1878) (Cumbria Record Office, Kendal, WDX/144).

56. George Henry Hill, "The Thirlmere Works for the Supply of Manchester," *Excerpt of Minutes of the Proceedings of the Institute of Civil Engineers* 126, pt. 4 (1895–96): 109.

57. R. D. Oldham, "Beach Formation in the Thirlmere Reservoir," *Journal of the Manchester Geographical Society* 16 (1900–1901): 225–26.

58. M.C.W.W. Thirlmere Scheme. Directions Given by Sub-Committee (September 19, 1878), 2–3; ibid. (April 21, 1879), 71 (Manchester Central Library, M231/Special S. C. Thirlmere, box 2).

59. Wilson, *Thirlmere to Manchester,* 7.

60. Roby X., "Thirlmere Past and Present. Part I," *North Lonsdale Magazine and Furness Miscellany* 4 (August 1900), 28.

61. H., "Manchester and Thirlmere" (letter to the editor), *Manchester Guardian* (February 1, 1911); H. D. Rawnsley, *A Coach Drive at the Lakes: Windermere to Keswick; and the Buttermere Round* (Keswick: T. Bakewell, 1902), 70.

62. "Tree-Felling at Thirlmere," *Manchester Guardian* (January 30, 1911).

63. Gordon G. Wordsworth, "Manchester and Thirlmere" (letter to the editor), *Manchester Guardian* (December 31, 1910).

64. T. Thornely, "Manchester v. Thirlmere," *Spectator* (February 25, 1911).

65. Many decades later, official Manchester publications continued to emphasize the combined functions—providing timber and countering erosion—of the Thirlmere plantations. *A Guide to Launchy Ghyll Forest Trail: Thirlmere,* an undated booklet published ca. 1970, also blames sheep for the "disappearance of the natural forest" and credits Manchester with restoring the tree cover.

66. Isaac Hinchliffe, "Thirlmere Again" (letter to the editor), *Manchester Guardian* (February 5, 1924).

67. G. A. K. Hervey, "Flowering Plants, Ferns and Mosses," in Hervey and Barnes, *Natural History,* 61–62; Tyler, *Thirlmere Mines,* 216–17; Bruce Thompson, *The Lake District and the National Trust* (Kendal: Titus Wilson, 1946), 16–17.

68. "Thirlmere Visited. Councillors Interesting Tour," [Manchester] *City News* (June 20, 1914); F. S. Arnold, "Thirlmere Again" (letter to the editor), *Manchester Guardian* (February 12, 1924).

69. *Times* (October 20, 1877), reprinted in Thirlmere Defence Association, *Extracts from the Leading Journals on the Manchester Water Scheme* (Windermere: J. Garnett, 1877), 5.

70. Farquhar, *Objections to the Thirlmere Scheme,* 21; Rawnsley, "Proposed Permanent Lake District Defense Society," 71.

71. Manchester was home to many working-class ramblers and naturalists, as described by Anne Secord in "Science in the Pub: Artisan Botanists in Early Nineteenth-Century Lancashire," *History of Science* 32 (1994): 269–315, and "Elizabeth Gaskell and the Artisan Naturalists of Manchester," *Gaskell Society Journal* 19 (2005): 34–51. On early ramblers, see Tom Stephenson, *Forbidden Land: The Struggle for Access to Mountain and Moorland* (Manchester: Manchester University Press, 1989), chap. 1.

72. Stephenson, *Forbidden Land,* 59.

73. *Closing of Public Footpaths: Indignation Meeting*... (Manchester: A. Ireland, 1866), 17 (in George Richard Jesse, "Public Rights of Way," British Library).

74. G. Shaw Lefevre, *English Commons and Forests: The Story of the Battle during the Last Thirty Years for Public Rights over the Commons and Forests of England and Wales* (London: Cassell, 1894), chaps. 2–3; W. H. Williams, *The Commons, Open Spaces, and Footpaths Preservation Society, 1865–1965: A Short History of the Society and Its Work* (London: Commons, Open Spaces, and Footpaths Preservation Society, 1965), 3–5.

75. Lefevre, *English Commons and Forests,* 338–39.

76. David Hollett, *The Pioneer Ramblers, 1850–1940* (Manchester: North Wales Area of the Ramblers' Association, 2002), 129.

77. William Bell, *Ancient Footpaths: The Latrigg Case* (Manchester, 1888), 1 (Armitt Library, LD1/52).

78. Geoffrey Berry and Geoffrey Beard, *The Lake District: A Century of Conservation* (Edinburgh: John Bartholomew, 1980), 3.

79. Marshall and Walton, *Lake Counties,* 218; Joseph Baron, *All about the English Lakes: A Cyclopedia of Places, Persons, Myths and Happenings* (Kendal: Atkinson and Pollitt, 1925), 116.

80. Marshall and Walton, *Lake Counties,* 218; Graham Murphy, *Founders of the National Trust* (London: National Trust Enterprises, 2002), 86–88.

81. Oddfellow, *King Croesus; or, Harlequin Town Clerk Irwell and the Thirlmere Lake* (London: Samuel French, 1878), 11–12.

82. "The Manchester Corporation as Path-grabbers" (letter to the editor from the secretary of the Kendal and District Footpath Preservation Association), *Pall Mall Gazette* (June 14, 1886).

83. Norman Hoyle and Kenneth Sankey, *Thirlmere Water: A Hundred Miles, A Hundred Years* (Bury, Lancs.: Centwrite, 1994), 72.

84. William T. Palmer, *The English Lakes* (London: Adam and Charles Black, 1925), 160.

85. Harriet Martineau, *Guide to Keswick and Its Environs* (Windermere: John Garnett, 1857), 7.

86. "The Thirlmere Water Scheme," *Manchester Courier* (October 31, 1877), 7, reprinted in Manchester Corporation Waterworks, *Supply of Water from Thirlmere. Extracts from the Manchester Newspapers* (Manchester: A. Ireland, 1877); James Mansergh, *The Thirlmere Water Scheme of the Manchester Corporation, with a Few Remarks on the Longdendale Works and Water Supply Generally* (London: E. and F. N. Spon, 1878), 35.

87. Farquhar, *Objections to the Thirlmere Scheme,* 12; "May 1848. Manchester Water Bill. Mr. Harward's Mem'd. For Mr. Easton as to opposition" (Manchester Central Library, M48, box 2).

88. Wilson, *Thirlmere to Manchester,* 43, 44.

89. *Protection of Thirlmere. Memorial of the Commons Preservation and Lake District Defence Societies to the Committee of the Manchester Water Works, Thirlmere* (flier, Armitt Library, LD 1/64).

90. "The Manchester Water Supply. The New Thirlmere Scheme," *Manchester Courant* (December 6, 1888).

91. J. B. Baddeley, "Thirlmere," *Manchester Guardian* (December 10, 1888).

92. W. H. Hills to G. Shaw Lefevre (January 26, 1889) (Cumbria Record Office, Carlisle).

93. W. H. Hills to Mr. Berrey, superintendent of the Waterworks Department (January 15, 1889) (Cumbria Record Office, Carlisle).

94. See, for example, "County Elector," *County Council Election. Keswick Division* (Keswick: N. V. Swindle, 1889), and W. H. Hills, *County Council Election. Keswick Division* (Keswick: T. Bakewell, 1889) (fliers in Cumbria Record Office, Carlisle).

95. *Manchester and Thirlmere. Substitution of a New Carriage Road for the Old Pack-horse Road* (Bowness-on-Windermere Association, 1889) (flier in Cumbria Record Office, Carlisle); "Cockermouth Union Highway Authority. The Thirlmere Roads," *West Cumberland Times* (April 10, 1889).

96. English Lake District Association, *Report of the Committee for the Year Ending 31st December 1919* (Armitt Library, LD1/59).

97. Wilson, *Thirlmere to Manchester,* 45; M. J. B. Baddeley, ed., *Black's Shilling Guide to the English Lakes* (London: Adam and Charles Black, 1903), 59.

98. Wilson, *Thirlmere to Manchester,* 10; H. D. Rawnsley, *Literary Associations of the English Lakes* (Glasgow: James MacLehose, 1894), 219. The bard in this quotation is Wordsworth.

99. House of Commons Select Committee on the Manchester Corporation Water Bill (Shorthand Writer's Notes), Minutes of Evidence (March 15, 1878), 306.

100. Harwood, *History and Description,* 40; Blunt, "The Translation of Thirlmere," 10.

101. Rawnsley, *Literary Associations,* 218; Thomas Huson, *Round about Helvellyn* (London: Seeley, 1895), 29. In 1984 the remains of the Rock of Names were moved to the Wordsworth Museum at Dove Cottage, near Grasmere. Grevel Lindop, *A Literary Guide to the Lake District* (London: Chatto and Windus, 1993), 123.

102. Arthur Redford, *The History of Local Government in Manchester,* vol. 2, *Borough and City* (London: Longmans, Green, 1940), 346–47.

103. Harwood, *History and Description,* 167–69.

104. Redford, *History of Local Government,* 2:371.

105. "The Thirlmere Waterworks. The Inaugural Ceremony," *Westmoreland Gazette* (October 20, 1894).

106. "Thirlmere Waterworks. Inaugural Ceremony in Cumberland," *Manchester Guardian* (October 13, 1894).

107. Harwood, *History and Description,* 173.

108. Ibid., 171, 177.

109. H. D. Rawnsley, *Commemorative Sonnets. To the Chairman and the Members of the Waterworks Committee of the Manchester Corporation, the Engineers, Builders, and Workmen, on the occasion of the Opening of the Thirlmere Waterworks, Friday, 12th October, 1894* (n.p., 1894).

110. "Thirlmere Waterworks," *Manchester Faces and Places* 6 (December 1894), 12; "Thirlmere Waterworks. Inaugural Ceremony in Cumberland," *Manchester Guardian* (October 13, 1894).

111. Harwood, *History and Description,* 179.

112. "The Opening of Thirlmere Waterworks. Ceremony in Manchester," *Manchester Guardian* (October 15, 1894). Triumphalism was a standard element of civic celebrations of new waterworks. See Owen G. Roberts, "Waterworks and Commemoration: Purity, Rurality, and Civic Identity in Britain, 1880–1921," *Continuity and Change* 22 (2007): 305–25.

113. Printed menus for the "Opening of the Thirlmere Waterworks" luncheon (October 12, 1894) and dinner (October 13, 1894) (Manchester Central Library, MSC 628/13).

114. "The Manchester-Thirlmere Waterworks," *Supplement to the Illustrated London News* (October 13, 1894).

115. G. Teymon, "La grand Digue du lac Thirlmère," *La Science Illustrée* 15, no. 366 (1894): 7; translated by the author.

116. New South Wales National Parks and Wildlife Service, *Thirlmere Lakes National Park: New Plan of Management* (November 1997), 17–18; personal communication from Marian Moore, Senior Librarian, NPWS Library, NSW.

117. Blunt, "Translation of Thirlmere," 7; Roby X., "Thirlmere, Past and Present. Part I," 27.

118. William T. Palmer and A. Heaton Cooper, *The English Lakes* (London: Adam and Charles Black, 1908), 173, 175; Mabel Howard, "Some Practical Notes on Cycling," *Badminton Magazine of Sports and Pastimes* (February 1898), 169; Mrs. Humphrey Ward, *Missing* (1917) (extract in scrapbook, "M.C.W.W. Newspaper Cuttings," Manchester Central Library, M231/125).

119. Teymon, "La grand Digue," 8.

120. *Historical Record,* 55, 57.

121. "Thirlmere Visited. Councillors Interesting Tour," *City News* (Manchester) (June 20, 1914).

122. *The English Lakes* (London: London and Northwestern Railway, 1910), 22.

123. Thomas D. Murphy, *In Unfamiliar England: A Record of a Seven Thousand Mile Tour by Motor*... (Boston: L. C. Page, 1910), 179.

124. Robert Shackleton, *Touring Great Britain* (Philadelphia: Penn Publishing, 1920), 315.

Chapter 5

1. "Re The Stewardship of the Manor of Legburthwaite in the County of Cumberland" (December 3, 1879) (manuscript, Cumbria Record Office, Carlisle, S/NWA/3/13).

2. Arthur Redford, *The History of Local Government in Manchester,* vol. 2, *Borough and City* (London: Longmans, Green, 1940), 348–51; John James Harwood, *The History and Description of the Thirlmere Water Scheme* (Manchester: Henry Blacklock, 1895), 45–46.

3. Court Book, Manor of Wythburn (1851–1928), 133 (manuscript, Cumbria Record Office, Carlisle, S/NWA/1/2); Manchester Corporation Waterworks, Manors of Legburthwaite and Wythburn. Appointment of Mr. J. Broatch as Steward, January 21, 1902 (manuscript, Cumbria Record Office, Carlisle, S/NWA/3/17).

4. Harwood, *History and Description,* vi.

5. "The Old Tenisonians A.D.C." *Era* (March 12, 1892).

6. Manchester Corporation Waterworks, *General Charges for Water within and beyond the Limits of the City* (Manchester: Henry Blacklock, 1908); Manchester Corporation Waterworks, *Thirlmere Scheme* (Manchester: Henry Blacklock, [1908?]) (Manchester Central Library, MSC 628).

7. "The Exhibition in the Making," *Manchester Guardian* (February 21, 1924). For a recent version, see the display about the Thirlmere Scheme at the Manchester Museum of Science and Industry.

8. H. D. Rawnsley, *A Coach Drive at the Lakes: Windermere to Keswick; and the Buttermere Round* (Keswick: T. Bakewell, 1902), 65.

9. Lake District Special Planning Board and North West Water Authority, *Thirlmere: Statement of Opportunities* (Kendal: Frank Peters, 1981), 17.

10. *Blackwell: The Arts and Crafts House* (Bowness on Windermere: Lakeland Arts Trust, 2002), 7–9.

11. Margaret Armstrong, ed., *Thirlmere across the Bridges to Chapel, 1849–1852: From the Diary of Reverend Basil R. Lawson, Curate of Wythburn* (Keswick: Peel Wyke, 1989), 33; T. Bulmer, *History, Topography, and Directory of Cumberland* (Preston: T. Snape, 1901), 484.

12. H. H. Symonds, *Walking in the Lake District* (London: Alexander Maclehose, 1935), 22.

13. "Thirlmere Waterworks," *Manchester Faces and Places* 6 (December 1894), 41.

14. Manchester Corporation Waterworks, *1847–1974. Chronology* (Manchester, [1974?]), 4–5 (pamphlet, Manchester Central Library).

15. "Manchester Corporation and Thirlmere. Councillor D. J. Mason Criticises the Committee. Councillor Baines' Defence," *West Cumberland Times* (March 8, 1924); "Keswick Floods. Thirlmere Waterworks Blamed," *Manchester Guardian* (January 10, 1925).

16. Harwards, Shepherd and Mills, "Manchester Water Bill. Instructions for Cross Examination of Mr. Bateman," 5 (Manchester Central Library, M48, box 11).

17. John Frederic La Trobe Bateman, *History and Description of the Manchester Waterworks* (Manchester: T. J. Day, 1884), 217.

18. J. H. Balfour Browne, *Water Supply* (London: Macmillan, 1880), 14.

19. M. F. Kennard, C. L. Owens, and R. A. Reader, *Engineering Guide to the Safety of Concrete and Masonry Dam Structures in the UK* (London: Construction Industry Research and Information Association, 1996), 67.

20. R. F. Critchley and D. J. Aikman, "Aqueduct Management Planning: Thirlmere, Haweswater, and Vyrnwy Aqueducts," *Journal of the Institution of Water and Environmental Management* 8 (1994): 507.

21. "Bursting of the Thirlmere Water Main," *Times* (July 5, 1899), 9; "Havoc of Thirlmere Main Pipeline Burst," *Evening Standard* (December 30, 1935).

22. "Dastardly Plot. Thirlmere Water Pipe in Danger," *Evening News* (April 5, 1921).

23. "Air Peril to Water Supply is 'Remote,'" *Evening Chronicle* (July 1937) (scrapbook, Manchester Central Library).

24. Isaac Hinchliffe, "Thirlmere Again" (letter to the editor), *Manchester Guardian*

(February 5, 1924); John W. Graham, "Thirlmere Again" (letter to the editor), *Manchester Guardian* (February 6, 1924).

25. "The Woods at Thirlmere. Viscount Ullswater's Hint to Manchester," *Manchester City News* (August 29, 1925).

26. Typescript notes of meeting between representatives of the Lake District Defense Society and the Manchester Corporation Waterworks Committee (November 12, 1926) (Cumbria Record Office, Carlisle).

27. Symonds, *Walking in the Lake District,* 83; H. H. Symonds, "Thirlmere, 1947" (typescript notes, Cumbria Record Office, Carlisle, DX/68/61); City of Manchester, *Centenary of the Waterworks Undertaking 1847–1947* (Manchester, 1947), 28.

28. "Vistas of Thirlmere," *Manchester Guardian* (October 9, 1959); "Thirlmere's Beauty Revealed," *Manchester Guardian* (January 6, 1960).

29. In the southern part of the Lake District, deciduous coppice woods designed to provide fuel for iron furnaces formed a traditional part of the landscape; they were at their largest in the eighteenth and nineteenth centuries. Roy Millward and Adrian Robinson, *The Lake District* (London: Eyre and Spottiswoode, 1970), 82, 85.

30. Forestry Commission, *Afforestation in the Lake District: Report by the Joint Informal Committee of the Forestry Commission and the Council for the Preservation of Rural England* (London: HMSO, 1936), 3.

31. G. A. K. Hervey, "Flowering Plants, Ferns and Mosses," in G. A. K. Hervey and J. A. G. Barnes, eds., *Natural History of the Lake District* (London: Frederick Warne, 1970), 61.

32. On this debate, see William Rollinson, *A History of Man in the Lake District* (London: J. M. Dent, 1967), 148–51; Geoffrey Berry and Geoffrey Beard, *The Lake District: A Century of Conservation* (Edinburgh: John Bartholomew, 1980), 14–19; John Sheail, *An Environmental History of Twentieth-Century Britain* (Basingstoke: Palgrave, 2002), 94–98; and John Sheail, *Nature in Trust: The History of Nature Conservation in Britain* (Glasgow: Blackie, 1976), 83–84.

33. Forestry Commission, *Afforestation,* 3.

34. H. H. Symonds, *Afforestation in the Lake District: A Reply to the Forestry Commission's White Paper of 26th August 1936* (London: J. M. Dent, 1936); "Trees in the Lake District, Forestry Commission Indicted," *Liverpool Daily Post* (December 29, 1936); E. Royce, "Afforestation in the Lake District," *Northern Rambler* 11 (January 1937).

35. Symonds, *Afforestation in the Lake District,* xi.

36. H. L. Edlin, *England's Forests: A Survey of the Woodlands Old and New in the English and Welsh Counties* (London: Faber and Faber, 1958), 212–13.

37. Bruce L. Thompson, *The Lake District and the National Trust* (Kendal: Titus Wilson, 1946), 16; Alfred Wainwright, *A Pictorial Guide to the Lakeland Fells,* bk. 3, *The Central Fells* (1958; London: Michael Joseph, 1992), Armboth Fell 3 (pagination in this book is by section).

38. George Bott, *Keswick: The Story of a Lake District Town* (Keswick: Cumbria County Library, 1994), 104.

39. Forestry Commission, "Forestry Commission Ends 70-Year-Old Dispute with Friends of the Lake District," News Release no. 7777 (June 30, 2005), http://www.forestry.gov.uk/newsrele.nsf/WebPressReleases/B32D14CE8EEECA818025702D00502B5C.

40. Geoffrey Halliday, "Changes in Cumbria's Flora in the Twentieth Century," in *Cumbrian Wildlife in the Twentieth Century,* ed. David J. Clarke and Stephen M. Hewitt,

Transactions of the Carlisle Natural History Society 12 (Carlisle: Carlisle Natural History Society, 1996), 9.

41. J. G. Baker, *A Flora of the English Lake District* (London: George Bell, 1885), 1–2; David Elliston Allen, *The Botanists: A History of the Botanical Society of the British Isles through a Hundred and Fifty Years* (Winchester: St Paul's Bibliographies, 1986), 34–35. The classes of citizenship, along with the standard "botanical provinces" of Britain, were established by Hewett Cottrell Watson in a series of publications in the middle decades of the nineteenth century. Albert Wilson, *The Flora of Westmorland* (Arbroath: T. Buncle, 1938), 58–60.

42. William Hodgson, *Flora of Cumberland* (Carlisle: W. Meals, 1898), x.

43. Baker, *Flora,* 2.

44. Ibid., 2–3.

45. Ibid., iv.

46. Geoffrey Halliday, *A Flora of Cumbria* (Lancaster: Centre for North-West Regional Studies, University of Lancaster, 1997), 70–74.

47. Derek Ratliffe, "Wildlife and Its Conservation in Cumbria," in Clarke and Hewitt, *Cumbrian Wildlife,* 96–97.

48. "Epilogue," in Hervey and Barnes, *Natural History,* 196–97.

49. H. A. Macpherson, *A Vertebrate Fauna of Lakeland, Including Cumberland and Wesmorland with Lancashire North of the Sands* (Edinburgh: David Douglas, 1892), 9, 44, 25, 27, 35.

50. *The Thirlmere Water Scheme. Reprinted from the Manchester Guardian, October 31st, 1877* (Manchester: J. Ireland, 1877), 4.

51. John Grave, "Country Notes. The Wild Birds of Thirlmere. IV. Destruction of Rare Birds: The Snow Bird and Golden Wren," *Manchester City News* (September 13, 1884); letter from Harold G. Waters, Waterworks Committee secretary, to T. T. Pickup (March 9, 1939) (Cumbria Record Office, Carlisle, DX68/35).

52. "The Screening of Thirlmere," *Manchester Guardian* (July 27, 1927).

53. "Rusticus," "By Hill, Dale, Wood, and Stream," *Lancashire Daily Post* (May 3, 1919).

54. W. T. Palmer, *The English Lakes* (London: Adam and Charles Black, 1943), 158; P. Delap, "Mammals," in Hervey and Barnes, *Natural History,* 186–87; Derek Yalden, *The History of British Mammals* (London: T. and A. D. Poyser, 1999), 172.

55. "Death of Mr. Isaac Thompson," *Cumberland and Westmorland Herald* (September 9, 1938); R. B. Graham, "Water Authorities and Farming: The Case of Thirlmere," *Manchester Guardian* (July 17, 1940).

56. Letter from J. P. Hallam, Waterworks Committee Engineer and Manager, to Rev. H. H. Symonds (July 31, 1940) (Cumbria Record Office, Carlisle, DX68/35).

57. Hervey, "Flowering Plants," in Hervey and Barnes, *Natural History,* 37–45; Ratliffe, "Wildlife and Its Conservation in Cumbria," 99; Baker, *Flora,* 5.

58. John Edward Marr, *Cumberland* (Cambridge: Cambridge University Press, 1910), 61.

59. Peder Anker, *Imperial Ecology: Environmental Order in the British Empire, 1895–1945* (Cambridge, MA: Harvard University Press, 2001), 18–19.

60. For a detailed account, see Robert W. Righter, *The Battle over Hetch Hetchy: America's Most Controversial Dam and the Birth of Modern Environmentalism* (New York: Oxford University Press, 2005).

61. M. M. O'Shaughnessy, *Hetch Hetchy: Its Origin and History* (San Francisco: n.p., 1934), 11.

62. The Sierra Club had been founded in 1892 to safeguard Yosemite. Holway R. Jones, *John Muir and the Sierra Club: The Battle for Yosemite* (San Francisco: Sierra Club, 1965), 11.

63. Letter from John Ripley Freeman to Marsden Mansen, San Francisco city engineer (January 27, 1911) (MIT Archives and Special Collections, Freeman Papers).

64. Letter from John Ripley Freeman to Marsden Mansen, San Francisco city engineer (September 3, 1910) (MIT Archives and Special Collections, Freeman Papers).

65. Letter from John Ripley Freeman to Marsden Mansen, San Francisco city engineer (September 3, 1910) (MIT Archives and Special Collections, Freeman Papers).

66. Warren D. Hanson, *San Francisco Water and Power: A History of the Municipal Water Department and Hetch Hetchy System* (San Francisco: City and County of San Francisco, 1985), 22.

67. John R. Freeman, *On the Proposed Use of a Portion of the Hetch Hetchy, Eleanor and Cherry Valleys . . . as Reservoirs for Impounding Tuolumne River Flood Waters and Appurtenant Works for the Water Supply of San Francisco, California, and Neighboring Cities* (San Francisco: Board of Supervisors, 1912), 46–47.

68. Ibid., 52.

69. Elizabeth Porter, *Water Management in England and Wales* (Cambridge: Cambridge University Press, 1978), 39.

70. E. D. Simon, *A City Council from Within* (London: Longmans, Green, 1926), 184.

71. Shena D. Simon, *A Century of City Government: Manchester 1838–1938* (London: George Allen and Unwin, 1938), 356.

72. Sheail, *Environmental History,* 73.

73. T. C. Smout, *Nature Contested: Environmental History in Scotland and Northern England since 1600* (Edinburgh: Edinburgh University Press, 2000), 112; Berry and Beard, *Lake District,* 9.

74. *The English Lakes in the Neighbourhood of Keswick and Penrith* (London: T. Nelson, 1870), 13–14.

75. *The English Lake District, including Furness Abbey, Seascale, etc.,* Concise Series of Guides, no. 2 (Edinburgh: John Menzies, n.d.), 119.

76. Gordon Home, *What to See in the English Lakes* (London: A. and C. Black, 1925), 2, 28; W. T. Palmer, "Shepherds' Meets (Mardale, Westmorland)," *Fortnightly Review* (November 1925), 720; Palmer, *English Lakes,* 168–69.

77. Rollinson, *History of Man,* 152–53.

78. Joseph Baron, *All about the English Lakes: A Cyclopedia of Places, Persons, Myths and Happenings* (Kendal: Atkinson and Pollitt, 1925), 98. The lake has also served as the setting for the prizewinning novel *Haweswater* by Sarah Hall (London: Faber and Faber, 2002).

79. Geoffrey Berry, *Mardale Revisited: The Story of Haweswater* (Kendal: Westmorland Gazette, 1996), 9.

80. "The Octopus: Its Latest Extension into Lakeland," *Sphere* (June 25, 1921), 322; "Haweswater—A Doomed Lake," *Times* (January 29, 1919).

81. "Westmorland Water," *Westmorland Gazette* (February 8, 1919); "Waste!," *Daily Dispatch* (February 11, 1919).

82. "Notes and Comments," *Herald* (June 8, 1918).

83. Isaac Hinchliffe, *A Backwater in Lakeland: A War-time Wandering, an Autumn Pilgrimmage* (Manchester: Hinchliffe, 1928), 5, 62, 61.

84. "The Future of Haweswater. Civic Visit of the Manchester Corporation," *Penrith Observer* (July 21, 1925).

85. Berry, *Mardale Revisited,* 34.

86. Symonds, *Walking in the Lake District,* 8–9.

87. Porter, *Water Management,* 39; Berry and Beard, *Lake District,* 9.

88. Berry, *Mardale Revisited,* 9–10.

89. Porter, *Water Management,* 40–41; John Hassan, *A History of Water in Modern England and Wales* (Manchester: Manchester University Press, 1998), 101.

90. Porter, *Water Management,* 40.

91. Ibid., 41–43; Berry and Beard, *Lake District,* 10–11.

92. Hassan, *History of Water,* 77, 85–86.

93. United Utilities, "Water," http://www.unitedutilities.com/?OBH=4228.

94. Graham Murphy, *Founders of the National Trust* (London: National Trust, 2002), 102–5; Thompson, *Lake District,* 40–45.

95. H. D. Rawnsley, *By Fell and Dale at the English Lakes* (Glasgow: James MacLehose, 1911), 208; Rollinson, *History of Man,* 142.

96. William Wordsworth, *The Illustrated Wordsworth's Guide to the Lakes,* ed. Peter Bicknell (New York: Congdon and Weed, 1984), 133.

97. Berry and Beard, *Lake District,* 4–5. For a detailed account, see F. R. Sandbach, "The Early Campaign for a National Park in the Lake District," *Transactions of the Institute of British Geographers* 3 (1978): 498–514.

98. City of Manchester, *Centenary of the Waterworks,* 32, 28.

99. Martyn Halsall, "How 96-Mile Canal Gave Manchester the Best Cup of Tea," *Guardian* (October 13, 1994), 5.

100. "Thirlmere's Centenary Party," *Lake District Herald* (October 15, 1994), in *Thirlmere 100: Media Coverage* (n.p.: North West Water, 1994).

101. Charles Graham, "Vintage Year for a Toast to the Lakes," *Wigan Evening Post* (October 13, 1994); "Water Great Achievement!," *Bolton Evening News* (September 30, 1994). Both in *Thirlmere 100: Media Coverage.*

102. Simon Wolstencroft, "Dirty Old Town's Source of Purity," *East Manchester Reporter* (October 13, 1994), in *Thirlmere 100: Media Coverage.*

103. Desmond Pitcher, preface to Norman Hoyle and Kenneth Sankey, *Thirlmere Water: A Hundred Miles, A Hundred Years* (Bury, Lancs.: Centwrite, 1994), 6.

104. Hoyle and Sankey, *Thirlmere Water,* 73.

105. Wainwright, *Pictorial Guide,* Armboth Fell 2.

106. Philip Johnston, "Water Bill Step to Protect Landscape," *Daily Telegraph* (March 31, 1989), 1.

107. United Utilities, "Map of UU Recreation Sites," http://www.unitedutilities.com/resources/files/1342_Map of UU Recreation Sites.pdf.

Bibliography

Abraham, Ashley P. *Beautiful Lakeland.* Keswick: G. P. Abraham, 1912.
Allen, David Elliston. *The Botanists: A History of the Botanical Society of the British Isles through a Hundred and Fifty Years.* Winchester: St Paul's Bibliographies, 1986.
Anker, Peder. *Imperial Ecology: Environmental Order in the British Empire, 1895–1945.* Cambridge, MA: Harvard University Press, 2001.
Archaeology UK. "The Archaeological Sites Index (ARCHI)." http://www.digital-documents.co.uk/archi/archi.htm (accessed 11 September 2008).
Armstrong, Margaret, ed. *Thirlmere across the Bridges to Chapel, 1849–1852: From the Diary of Basil R. Lawson, Curate of Wythburn.* Keswick: Peele Wyke, 1989.
Aston, Joseph. *The Manchester Guide: A Brief Historical Description of the Towns of Manchester and Salford, the Public Buildings, and the Charitable and Literary Institutions.* Manchester: Joseph Aston, 1804.
Axon, William E. A. *The Annals of Manchester: A Chronological Record from the Earliest Times to the End of 1855.* Manchester: John Heywood, 1886.
Baddeley, M. J. B. *The English Lake District.* 6th ed. Thorough Guide Series. London: Dulau, 1891.
——, ed. *Black's Shilling Guide to the English Lakes.* London: Adam and Charles Black, 1903.

Bailey, John, and George Culley. *General View of the Agriculture of Northumberland, Cumberland and Westmorland.* 1805. Reprint, Newcastle-upon-Tyne: Frank Graham, 1972.

Baines, Edward. *A Companion to the Lakes of Cumberland, Westmoreland, and Lancashire.* London: Simpkin and Marshall, 1834.

Baker, J. G. *A Flora of the English Lake District.* London: George Bell, 1885.

Baldwin-Wiseman, William Ralph. "The Increase in the National Consumption of Water." *Journal of the Royal Statistical Society* 72 (1909): 243–303.

Balfour Browne, J. H. *Water Supply.* London: Macmillan, 1880.

Barber, Samuel. *Beneath Helvellyn's Shade: Notes and Sketches in the Valley of Wythburn.* London: Elliot Stock, 1892.

Barclay, Thomas. *The Future Water Supply of Birmingham.* Birmingham: Cornish Brothers, 1898.

Baron, Joseph. *All about the English Lakes: A Cyclopedia of Places, Persons, Myths and Happenings.* Kendal: Atkinson and Pollitt, 1925.

Bateman, John Frederic LaTrobe. *History and Description of the Manchester Waterworks.* Manchester: T. J. Day, 1884.

———. *Reports on the Various Schemes for Supplying Glasgow with Water.* Manchester: Cave and Sever, 1853.

Baynes, John and George Westropp. *Lake Vyrnwy: The Story of a Sporting Hotel.* Oswestry: Kettering Books, 1992.

Beardmore, Nathaniel. *Manual of Hydrology.* London: Waterlow and Sons, 1862.

Bell, William. *Ancient Footpaths: The Latrigg Case.* Manchester, 1888.

Berry, Geoffrey. *Mardale Revisited: The Story of Haweswater.* Kendal: Westmorland Gazette, 1996.

Berry, Geoffrey, and Geoffrey Beard. *The Lake District: A Century of Conservation.* Edinburgh: John Bartholomew, 1980.

Blackwell: The Arts and Crafts House. Bowness on Windermere: Lakeland Arts Trust, 2002.

Blanchard, E. L. *Adams's Pocket Guide to the Lake District of Lancashire, Westmoreland, and Cumberland.* London: W. J. Adams, 1852.

Bonney, Thomas George, and Elijah Walton. *English Lake Scenery.* London: W. M. Thompson, 1875.

Bott, George. *Keswick: The Story of a Lake District Town.* Keswick: Cumbria County Library, 1994.

Bouch, C. M. L., and G. P. Jones. *A Short Economic and Social History of the Lake Counties, 1500–1830.* Manchester: Manchester University Press, 1961.

Brabant, F. G. *The English Lakes.* London: Methuen, 1902.

Bradshaw, L. D., ed. *Visitors to Manchester: A Selection of British and Foreign Visitors' Descriptions of Manchester from c1538 to 1865.* Manchester: Neil Richardson, 1987.

Braithwaite, George Foster. *The Salmonidae of Westmorland, Angling Reminiscences, and Leaves from an Angler's Notebook.* Kendal: Atkinson and Pollitt, 1884.

Breeze, Laurence E. *The British Experience with River Pollution, 1865–1876.* New York: Peter Lang, 1993.

Brierley, Ben. *A Trip to Thirlmere and Borrowdale.* Manchester: Abel Heywood, 1878.

Briggs, Asa. *Victorian Cities.* New York: Harper and Row, 1970.

Budworth, Joseph [Joseph Palmer]. *A Fortnight's Ramble to the Lakes in Westmoreland, Lancashire and Cumberland. By a Rambler.* London: Hookham and Carpenter, 1792.

Bulmer, T. *History, Topography, and Directory of Cumberland.* Preston: T. Snape, 1901.

Buzard, James. *The Beaten Track: European Tourism, Literature and the Ways to "Culture," 1800–1918.* Oxford: Oxford University Press, 1993.

Cameron, Alastair, ed. *Lakeland's Mining Heritage: The Last 500 Years.* Alston, Cumbria: Cumbrian Amenity Trust Mining History Society, 2000.

Capelli, Tim. *The Thirlmere Way: A Long Distance Walk from Manchester to the Lake District.* Wilmslow, Cheshire: Sigma Press, 1992.

Carlyle, Thomas. *Past and Present.* 1843. Reprint, London: Dent, 1960.

City of Manchester. *Centenary of the Waterworks Undertaking 1847–1947.* Manchester: City of Manchester, 1947.

Clark, Geoffrey, and W. Harding Thompson. *The Lakeland Landscape.* London: Adam and Charles Black, 1938.

Clarke, David J., and Stephen M. Hewitt, eds. *Cumbrian Wildlife in the Twentieth Century.* Transactions of the Carlisle Natural History Society 12. Carlisle: Carlisle Natural History Society, 1996.

Clarke, James. *A Survey of the Lakes of Cumberland, Westmorland, and Lancashire.* London: James Clarke, 1787.

Clarke, John S. *An Epic of Municipalisation: The Story of Glasgow's Loch Katrine Water Supply.* Glasgow: Forward, 1928.

Clifford, Frederick. *A History of Private Bill Legislation.* Vols. 1 and 2. London: Butterworths, 1885.

Collingwood, W. G. *Lake Counties.* New York: E. P. Dutton, 1902.

———. *Lake District History.* Kendal: Wilson and Son, 1925.

Credat Judaeus [pseud.]. *Thirlmere Water Scheme. Remarks on the Evidence Given before the Select Committee of the House of Commons.* London: P. S. King, 1878.

Critchley, R. F., and D. J. Aikman. "Aqueduct Management Planning: Thirlmere, Haweswater, and Vyrnwy Aqueducts." *Journal of the Institution of Water and Environmental Management* 8 (1994): 502–12.

Cunliffe, Barry, Robert Bartlett, Joanna Bourke, Asa Briggs, and John Morrill, eds. *Penguin Atlas of British and Irish History.* London: Penguin, 2001.

Darby, H. C., ed. *A New Historical Geography of England.* Cambridge: Cambridge University Press: 1973.

Daunton, Martin. *Progress and Poverty: An Economic and Social History of Britain, 1700–1850.* Oxford: Oxford University Press, 1995.

Davis, R. V. *Geology of Cumbria: Lakeland's Rocks and Minerals Explained.* Clapham, N. Yorks.: Dalesman Books, 1977.

Davy, John. *The Angler in the Lake District; or, Piscatory Colloquies and Fishing Excursions in Westmoreland and Cumberland.* London: Longman, Brown, Green, Longmans and Roberts, 1857.

de Tocqueville, Alexis. *Journeys to England and Ireland.* Excerpted in Bradshaw, *Visitors to Manchester,* 32–35.

Defoe, Daniel. *A Tour through the Island of Great Britain.* Vol. 3. 1726. Reprint, London: J. F. and C. Rivington, 1778.

———. *A Tour through the Whole Island of Great Britain.* 1724–26. Reprint, edited by P. N. Furband and W. R. Owens. New Haven: Yale University Press, 1991.

De Quincey, Thomas. *Literary Reminiscences from the Autobiography of an English Opium Eater.* In *The Works of Thomas De Quincey,* vol. 3. Boston: Houghton Mifflin, 1851.

Dickson, Nicholas, ed. *The Queen's Guard of Honour, Loch Katrine, 1859. A Memorial Record.* Glasgow: N. Macphail, 1885.

Duffield, H. G. *The Strangers Guide to Manchester.* 1850. Reprint, Swinton: Neil Richardson, 1984.

Edlin, H. L. *England's Forests: A Survey of the Woodlands Old and New in the English and Welsh Counties.* London: Faber and Faber, 1958.

Eighteenth Annual Report of the Inspectors of Salmon Fisheries (England and Wales). (For the Year 1878). London: HMSO, 1879.

Engels, Friedrich. *The Condition of the Working Class in England.* 1845. Reprint, edited by David McLellan and translated by Florence Kelley-Wischnewetsky. Oxford: Oxford University Press, 1993.

The English Lake District. The Concise Series of Guides, no. 2. Kendal: Titus Wilson, 1908.

The English Lake District, including Furness Abbey, Seascale, etc. The Concise Series of Guides, no. 2. Edinburgh: John Menzies, n.d.

The English Lakes. London: London and Northwestern Railway, 1910.

The English Lakes in the Neighbourhood of Keswick and Penrith. London: T. Nelson, 1870.

Farquhar, Robert T. *Objections to the Thirlmere Scheme (Manchester Water Bill. 1879).* Ambleside: "Lakes Chronicle" Office, 1879.

Fell, Clare. *Early Settlement in the Lake Counties.* Clapham, Yorks.: Dalesman Books, 1972.

Ferguson's Tourists' Guide to Callander, the Trosachs, Loch Katrine, Loch Lomond . . . 9th ed. Glasgow: Thomas Murray, 1911.

Fiennes, Celia. *The Journeys of Celia Fiennes.* Edited by Christopher Morris. London: Cresset Press, 1949.

Forestry Commission. *Afforestation in the Lake District: Report by the Joint Informal Committee of the Forestry Commission and the Council for the Preservation of Rural England.* London: HMSO, 1936.

———. "Forestry Commission Ends 70-Year-Old Dispute with Friends of the Lake District." Forestry Commission News Release No. 7777, 30 June 2005. http://www.forestry.gov.uk/newsrele.nsf/ WebPressReleases/B32D14CE8EEECA818025702D00502B5C (accessed 11 September 2008).

Fraser, Derek, ed. *Municipal Reform and the Industrial City.* Leicester: Leicester University Press, 1982.

Freeman, John R. *On the Proposed Use of a Portion of the Hetch Hetchy, Eleanor and Cherry Valleys . . . as Reservoirs for Impounding Tuolumne River Flood Waters and Appurtenant Works for the Water Supply of San Francisco, California, and Neighboring Cities.* San Francisco: Board of Supervisors, 1912.

Gale, James M. *Papers Read before the Institution of Engineers in Scotland on the Glasgow Waterworks.* Glasgow: Gardner and Stevenson, 1864.

Gallichan, Walter M. [Geoffrey Mortimer, pseud.]. *Fishing in Wales: A Guide to the Angler.* London: F. E. Robinson, 1903.

Gilpin, William. *Observations Relative Chiefly to Picturesque Beauty, Made in the Year 1772, on Several Parts of England; Particularly the Mountains, and Lakes of Cumberland and Westmoreland,* Vol. 1. 1786. Reprint, Poole, NY: Woodstock Books, 1996.

Gow, Bill. *The Swirl of the Pipes: A History of Water and Sewerage in Strathclyde.* Glasgow: Strathclyde Regional Council, 1996.

Gray, Thomas. *Thomas Gray's Journal of His Visit to the Lake District in October 1769.* Edited by William Roberts. Liverpool: Liverpool University Press, 2001.

Green, William. *A Description of Sixty Studies from Nature.* London: Longman, Hurst, Rees, and Orme, 1810.

Grisdale, John Wilson. *"A Voice from the Ranks": An Original Poem on Thirlmere Lake.* Keswick: Bakewell and Fitzjames, 1877.

A Guide to Launchy Ghyll Forest Trail: Thirlmere. n.d. [1970?].

Guide to Manchester and Salford. Edinburgh: Adam and Charles Black, 1882.

Hall, Sarah. *Haweswater.* London: Faber and Faber, 2002.

Halliday, Geoffrey. *A Flora of Cumbria.* Lancaster: Centre for North-West Regional Studies, Lancaster University, 1997.

Halliday, Stephen. *The Great Stink of London: Sir Joseph Bazalgette and the Cleansing of the Victorian Metropolis.* Stroud: Sutton Publishing, 1999.

Halsall, Martyn. "How 96-Mile Canal Gave Manchester the Best Cup of Tea." *Guardian,* 13 October 1994, 5.

Hamlin, Christopher. *A Science of Impurity: Water Analysis in Nineteenth-Century Britain.* Berkeley: University of California Press, 1990.

Hanson, Warren D. *San Francisco Water and Power: A History of the Municipal Water Department and Hetch Hetchy System.* San Francisco: City and County of San Francisco, 1985.

Hartwell, Clare. *Manchester.* London: Penguin Books, 2001.

Harwood, John James. *A Short Description of the Manchester Corporation Waterworks.* Manchester: H. Blacklock, 1891.

——. *History and Description of the Thirlmere Water Scheme.* Manchester: Henry Blacklock, 1895.

Hassan, John A. "The Growth and Impact of the British Water Industry in the Nineteenth Century." *Economic History Review* 38 (1985): 531–42.

——. *A History of Water in Modern England and Wales.* Manchester: Manchester University Press, 1998.

——, and E. R. Wilson. "The Longdendale Water Scheme 1848–1884." *Industrial Archaeology* 14 (1979): 102–21.

Hervey, G. A. K., ed. *Natural History of the Lake District.* London: Frederick Warne, 1970.

Hey, David. "Moorlands." In *The English Rural Landscape,* ed. Joan Thirsk, 138–209. Oxford: Oxford University Press, 2000.

Hibbert, F. "A Description of the Liverpool Corporation Water Supply Undertaking—1847–1947." *Water and Water Engineering* 51 (May 1948): 202–11.

Hill, George Henry. "The Thirlmere Works for the Supply of Manchester." *Excerpt of Minutes of the Proceedings of the Institute of Civil Engineers* 126, pt. 4 (1895–96).

Hilton, Tim. *John Ruskin: The Later Years.* New Haven: Yale University Press, 2000.

Hinchliffe, Isaac. *A Backwater in Lakeland: A War-time Wandering, an Autumn Pilgrimage.* Manchester: Hinchliffe, 1928.

An Historical Record of Some Recent Enterprises of the Corporation of Manchester and of Its Co-operation in the Completion of the Manchester Ship Canal. Manchester: Henry Blacklock, 1894.

Hodgson, William. *Flora of Cumberland.* Carlisle: W. Meals, 1898.

Hollett, David. *The Pioneer Ramblers, 1850–1940.* Manchester: North Wales Area of the Ramblers' Association, 2002.

Home, Gordon. *What to See in the English Lakes.* London: A. and C. Black, 1925.

House of Commons Select Committee on the Manchester Corporation Water Bill (Shorthand Writer's Notes). 1878.

Hoyle, Norman, and Kenneth Sankey. *Thirlmere Water: A Hundred Miles, A Hundred Years.* Bury, Lancs.: Centwrite, 1994.

Huddleston, C. Roy, and R. S. Boumphrey. *Cumberland Families and Heraldry.* n.p.: Cumberland and Westmorland Antiquarian and Archaeological Society, 1978.

Hull, Edward L. "The Geology." In *The Lake Counties,* edited by W. G. Collingwood, 243–49. New York: E. P. Dutton, 1902.

Huson, Thomas. *Round about Helvellyn.* London: Seeley, 1895.

Hutchinson, W. *An Excursion to the Lakes in Westmorland and Cumberland.* London: J. Wilkie, 1776.

Jenkinson, Henry Irwin. *Jenkinson's Practical Guide to the English Lake District.* 1st ed. London: E. Stanford, 1872.

Jenkinson, Henry Irwin. *Practical Guide to the English Lake District.* London: Edward Stanford, 1879.

Johnston, Kenneth R. *The Hidden Wordsworth: Poet, Lover, Rebel, Spy.* New York: W. W. Norton, 1998.

Johnston, Philip. "Water Bill Step to Protect Landscape." *Daily Telegraph,* 31 March 1989, 1.

Jones, Holway R. *John Muir and the Sierra Club: The Battle for Yosemite.* San Francisco: Sierra Club, 1965.

Kargon, Robert H. *Science in Victorian Manchester: Enterprise and Expertise.* Baltimore: Johns Hopkins University Press, 1977.

Kay, James Phillips. *The Moral and Physical Condition of the Working Classes Employed in the Cotton Manufacture in Manchester.* 1832. Facsimile reprint, Shannon: Irish University Press, 1971.

Kay, William. "Presidential Address." British Waterworks Association. 17th Annual Meeting, Manchester, 4–7 July 1928. n.p.: 1928.

Kendall, J. D. "Notes on the History of Mining in Cumberland and North Lancashire." *Transactions of the North of England Institute of Mining and Mechanical Engineers* 34 (1884–85): 83–124.

Kennard, M. F., C. L. Owens, and R. A. Reader. *Engineering Guide to the Safety of Concrete and Masonry Dam Structures in the UK.* London: Construction Industry Research and Information Association, 1996.

Kidd, Alan. "The Industrial City and Its Pre-industrial Past: The Manchester Royal Jubilee Exhibition of 1887." *Transactions of the Lancashire and Cheshire Antiquarian Society* 89 (1993): 63–70.

———. *Manchester.* Keele: Keele University Press, 1996.

King, John. *Manchester and Its Waterworks. Where Are We? A Letter on the Present Position of the Question, and Especially on Its Financial Aspect, to His Fellow-Members of the Manchester City Council.* Manchester: Charles Sever, 1884.

Lake District Special Planning Board and North West Water Authority. *Thirlmere: Statement of Opportunities.* Kendal: Frank Peters, 1981.

Lefevre, G. Shaw. *English Commons and Forests: The Story of the Battle during the Last Thirty Years for Public Rights over the Commons and Forests of England and Wales.* London: Cassell, 1894.

Lindop, Grevel. *A Literary Guide to the Lake District.* London: Chatto and Windus, 1993.

Linton, E. Linn. *The Lake Country*. London: Smith, Elder, 1864.

Luckin, Bill. *Pollution and Control: A Social History of the Thames in the Nineteenth Century*. Bristol: Adam Hilger, 1986.

Macpherson, H. A. *A Vertebrate Fauna of Lakeland, Including Cumberland and Wesmorland with Lancashire North of the Sands*. Edinburgh: David Douglas, 1892.

Manchester and Thirlmere Water Scheme: The Case of the Thirlmere Defence Association. Windermere: J. Garnett, 1878.

Manchester City Council. *The Thirlmere Water Supply Scheme. Speeches of Mr. Alderman King and Mr. Alderman Curtis*. London: C. F. Roworth, 1878.

Manchester Corporation Waterworks. *1847–1974. Chronology*. Manchester, [1974?].

———. *General Charges for Water within and beyond the Limits of the City*. Manchester: Henry Blacklock, 1908.

———. *Supply of Water from Thirlmere. Extracts from the Manchester Newspapers*. Manchester: A. Ireland, 1877.

———. *Thirlmere Scheme*. Manchester: Henry Blacklock, [1908?].

Manchester Corporation Waterworks Act 1879. Thirlmere Aqueduct. Lord Lonsdales's Estates. Arbitrations between the Right Hon. James Lowther, M.P. and the Corporation of Manchester. Awards of Christopher Oakley, Esq. (Umpire Appointed by the Board of Trade). December 21st, 1894. Manchester: Henry Blacklock, 1894.

Mannix and William Whellan. *History, Gazetteer and Directory of Cumberland*. 1847. Reprint, Cumberland: Michael Moon, 1974.

Mansergh, James. *The Thirlmere Scheme of the Manchester Corporation, with a Few Remarks on the Longdendale Works and Water Supply Generally*. London: E and F. N. Spon, 1878.

Marr, John Edward. *Cumberland*. Cambridge: Cambridge University Press, 1910.

———. *The Geology of the Lake District, and the Scenery as Influenced by Geological Structures*. Cambridge: Cambridge University Press, 1916.

Marshall, J. D., and John K. Walton. *The Lake Counties from 1830 to the Mid-twentieth Century: A Study in Regional Change*. Manchester: Manchester University Press, 1981.

Martineau, Harriet. *Complete Guide to the English Lakes*. Windermere: J. Garnett, 1855.

———. *Guide to Keswick and Its Environs*. Windermere: John Garnett, 1857.

Mathis, Charles-François. *In Nature We Trust: Les paysages anglais à l'ére industrielle*. Paris: Presses Universitaires Paris-Sorbonne, 2009.

Maver, Irene. "Children and the Quest for Purity in the Nineteenth Century Scottish City." *Paedagogica Historica* 33 (1997): 801–24.

McFadzean, Alen. *Wythburn Mine and the Lead Miners of Helvellyn*. Ulverston, Cumbria: Red Earth Publications, 1987.

A Microscopical Examination of Certain Waters Submitted to Jabez Hogg, Surgeon to the Royal Westminster Ophthalmic Hospital . . . and a Chemical Analysis by Dugald Campbell, Analytical Chemist to the Brompton Hospital. London: Balliere, Tindall, Cox, 1874.

Miller, John Fletcher. "On the Meteorology of the Lake District of Cumberland and Westmoreland, Part I." *Philosophical Transactions of the Royal Society of London* (1849): 73–89, 319–29.

Millward, Roy, and Adrian Robinson. *The Lake District*. London: Eyre and Spottiswoode, 1970.

"The Mosley Family of Manchester." http://www.thornber.net/cheshire/htmlfiles/mosley.html (accessed 11 September 2008).

Murphy, Graham. *Founders of the National Trust.* London: National Trust Enterprises, 2002.

Murphy, Thomas D. *In Unfamiliar England: A Record of a Seven Thousand Mile Tour by Motor*... Boston: L. C. Page, 1910.

New South Wales National Parks and Wildlife Service. *Thirlmere Lakes National Park: New Plan of Management.* n.p.: 1997.

Nicolson, Joseph, and Richard Burn. *The History and Antiquities of the Counties of Westmorland and Cumberland.* Vol. 2. London: W. Strahan and T. Cadell, 1777.

Notes on the Water Supply of Glasgow, Prepared on the Occasion of the Celebration of the Jubilee of the Loch Katrine Water-Works, 14th October, 1909. Glasgow: Glasgow Corporation Water Department, 1909.

An Oddfellow [pseud.]. *King Croesus; or, Harlequin Town Clerk Irwell and the Thirlmere Lake (An Unacted Fairy Extravaganza).* London: Samuel French, 1878.

Ogden, James. *A Description of Manchester by a Native of the Town.* Edited by Neil Richardson. Manchester: Neil Richardson, 1983.

Oldham, R. D. "Beach Formation in the Thirlmere Reservoir." *Journal of the Manchester Geographical Society* 16 (1900–1901): 225–26.

Oldroyd, David. "Early Ideas about Glaciation in the English Lake District: The Problem of Making Sense of Glaciation in a Glaciated Region." *Annals of Science* 56 (1999): 175–203.

O'Shaughnessy, M. M. *Hetch Hetchy: Its Origin and History.* San Francisco: n.p., 1934.

Otley, Jonathan. *A Concise Description of the English Lakes and Adjacent Mountains.* Keswick: Jonathan Otley, 1825.

Ouida [Louise de la Ramée]. "The Ugliness of Modern Life." In *Critical Studies.* New York: Cassell, n.d.

Palmer, William T. *The English Lakes.* London: Adam and Charles Black, 1943.

Palmer, William T., and A. Heaton Cooper. *The English Lakes.* London: Adam and Charles Black, 1908.

Parkinson-Bailey, John J. *Manchester: An Architectural History.* Manchester: Manchester University Press, 2000.

Pearsall, W. H., and Winifred Pennington. *The Lake District: A Landscape History.* London: Collins, 1973.

Pevsner, Nikolaus. *Cumberland and Westmorland.* Harmondsworth, Middlesex: Penguin, 1967.

A Picturesque Tour of the English Lakes, Containing a Description of the Most Romantic Scenery of Cumberland, Westmoreland, and Lancashire. London: R. Ackerman, 1821.

Piggott, Stuart, ed. *Prehistory: The Agrarian History of England and Wales.* Vol. 1, pt. 1. Cambridge: Cambridge University Press, 1981.

Platt, Harold L. *Shock Cities: The Environmental Transformation and Reform of Manchester and Chicago.* Chicago: University of Chicago Press, 2005.

Pooley, Marilyn E., and Colin G. Pooley. "Health, Society and Environment in Victorian Manchester." In *Urban Disease and Mortality in Nineteenth-Century England,* edited by Robert Woods and John Woodward, 148–75. London: Batsford, 1984.

Porter, Dale H. *The Thames Embankment: Environment, Technology and Society in Victorian London.* Akron: University of Akron Press, 1998.

Porter, Elizabeth. *Water Management in England and Wales.* Cambridge: Cambridge University Press, 1978.

Prior, Herman. *Guide to the Lake District of England.* 3rd ed. Windermere: J. Garnett, n.d. [1870?].

Quayle, Tom. *The Cotton Industry in Longdendale and Glossopdale.* Stroud: Tempus, 2006.

———. *Reservoirs in the Hills: The Story of the Construction of the Reservoirs in the Longdendale Valley which Feed Manchester with Water.* n.p.: Senior Publications, 1988.

Rackham, Oliver. *Trees and Woodland in the British Landscape: The Complete History of Britain's Trees, Woods and Hedgerows.* London: Phoenix Press, 1990.

Radcliffe, Ann. *A Journey made in the Summer of 1794 . . . to which are added, Observations during a tour to the Lakes of Lanchashire, Westmoreland, and Cumberland.* Dublin: William Porter, 1795.

Rawnsley, H. D. *A Coach Drive at the Lakes: Windermere to Keswick; and the Buttermere Round.* Keswick: T. Bakewell, 1902.

———. *Commemorative Sonnets. To the Chairman and the Members of the Waterworks Committee of the Manchester Corporation, the Engineers, Builders, and Workmen, on the Occasion of the Opening of the Thirlmere Waterworks, Friday, 12th October, 1894.* n.p.: 1894.

———. *By Fell and Dale at the English Lakes.* Glasgow: James MacLehose, 1911.

———. *Literary Associations of the English Lakes,* vol. 2. Glasgow: James MacLehose, 1894.

———. *Past and Present at the English Lakes.* Glasgow: James MacLehose, 1916.

———. "The Proposed Permanent Lake District Defense Society." *Transactions of the Cumberland Association for the Advancement of Literature and Science* 7 (1882–83): 77–80.

Redford, Arthur. *The History of Local Government in Manchester.* Vol. 2, *Borough and City.* London: Longmans, Green, 1940.

Report from the Select Committee on the Manchester Corporation Water Bill; with the Proceedings of the Committee. 1878.

Richards, Jeffrey. "The Role of the Railways." In *Ruskin and Environment: The Storm Cloud of the Nineteenth Century,* edited by Michael Wheeler, 123–43. Manchester: Manchester University Press, 1995.

Righter, Robert W. *The Battle over Hetch Hetchy: America's Most Controversial Dam and the Birth of Modern Environmentalism.* New York: Oxford University Press, 2005.

Roberts, Owen G. "Waterworks and Commemoration: Purity, Rurality, and Civic Identity in Britain, 1880–1921." *Continuity and Change* 22 (2007): 305–25.

Robinson, John. *A Guide to the Lakes in Cumberland, Westmorland, and Lancashire.* London: Lackington, Hughes, Harding, Mavor, and Jones, 1819.

Rollinson, William. *A History of Man in the Lake District.* London: J. M. Dent, 1967.

Royal Commission on Water Supply. *Report of the Commissioners.* London: Spottiswoode, 1869.

Ruskin, John. *Fors Clavigera: Letters to the Workmen and Labourers of Great Britain,* Vol. 2, 1871–84. Reprint, Boston: Colonial Press, 1900.

———. *A Tour to the Lakes in Cumberland: John Ruskin's Diary for 1830.* Edited by James S. Dearden. Aldershot: Scolar Press, 1990.

———. *Works.* Edited by E. T. Cook and Alexander Wedderburn. Cambridge: Cambridge University Press, 1996.

Russell, Peter. "Bateman, John Frederic LaTrobe (1810–1889). In *Oxford Dictionary of National Biography,* edited by Lawrence Goldman. Oxford: Oxford University Press, 2004.

Sandbach, F. R. "The Early Campaign for a National Park in the Lake District." *Transactions of the Institute of British Geographers* 3 (1978): 498–514.

Scott, Joseph. *Leaves from the Diary of a Citizen Auditor.* Manchester: Office of the Manchester City News, 1894.

Secord, Anne. "Elizabeth Gaskell and the Artisan Naturalists of Manchester." *Gaskell Society Journal* 19 (2005): 34–51.

———. "Science in the Pub: Artisan Botanists in Early Nineteenth-Century Lancashire." *History of Science* 32 (1994): 269–315.

Shackleton, Robert. *Touring Great Britain.* Philadelphia: Penn Publishing, 1920.

Sheail, John. *An Environmental History of Twentieth-Century Britain.* Basingstoke: Palgrave, 2002.

———. *Nature in Trust: The History of Nature Conservation in Britain.* Glasgow: Blackie, 1976.

Simon, E. D. *A City Council from Within.* London: Longmans, Green, 1926.

Simon, Shena D. *A Century of City Government: Manchester 1838–1938.* London: George Allen and Unwin, 1938.

Smout, T. C. *Nature Contested: Environmental History in Scotland and Northern England since 1600.* Edinburgh: Edinburgh University Press, 2000.

Somervell, Robert. *Chapters of Autobiography.* London: Faber and Faber, 1935.

———. *A Protest against the Extension of Railways in the Lake District.* Windermere: J. Garnett, 1876.

SS Sir Walter Scott on Loch Katrine. Norwich: Jarrold Publishing, 1994.

Stephenson, Tom. *Forbidden Land: The Struggle for Access to Mountain and Moorland.* Manchester: Manchester University Press, 1989.

Sutherland, Douglas. *The Yellow Earl: The Life of Hugh Lowther, 5th Earl of Lonsdale, 1857–1944.* 1965; Exeter: Molendinar Press, 1980.

Symonds, H. H. *Afforestation in the Lake District: A Reply to the Forestry Commission's White Paper of 26th August 1936.* London: J. M. Dent, 1936.

———. *Walking in the Lake District.* London: Alexander Maclehose, 1935.

Taggart, Michael. *Private Property and Abuse of Rights in Victorian England: The Story of Edward Pickles and the Bradford Water Supply.* Oxford: Oxford University Press, 2002.

Thirlmere Defence Association. *Extracts from the Leading Journals on the Manchester Water Scheme.* Windermere: J. Garnett, 1878.

———. *Is Thirlmere the Only Source of Supply for Manchester?* London: Sir Joseph Causton, 1878.

———. *Report to the Subscribers to the above Association upon the Manchester Corporation Water Works Bill for obtaining a Supply of Water from Lake Thirlmere.* n.p.: 1878.

Thirlmere Defence Association. The Case Re-stated. Windermere: J. Garnett, 1878.

Thirlmere 100: Media Coverage. n.p: North West Water, 1994.

The Thirlmere Water Scheme. Reprinted from the Manchester Guardian, October 31st, 1877. Manchester: J. Ireland, 1877.

Thompson, Bruce L. *The Lake District and the National Trust.* Kendal: Titus Wilson, 1946.

———. "The Leathes Family of Dalehead." *Transactions of the Cumberland and Westmorland Antiquarian and Archaeological Society,* n.s., 60 (1960): 109–19.

Thresh, John C. *Water and Water Supplies.* 1896. Reprint, Philadelphia: P. Blakiston's Son, 1901.

26th, 27th, and 28th September, 1894. The Manchester Corporation Waterworks Act 1879. Thirlmere Aqueduct. Lord Lonsdale's Estates. Arbitration between the Right Hon. James

Lowther, M. P. and the Corporation of Manchester. Before Christopher Oakley, Esq. Umpire Appointed by the Board of Trade. Manchester: Henry Blacklock, 1894.

Tyler, Ian. *Thirlmere Mines and the Drowning of the Valley.* Keswick: Blue Rock Publications, 1999.

United Utilities. "Water." http://www.unitedutilities.com/?OBH=4228 (accessed 11 September 2008).

———. "Map of UU Recreation Sites." http://www.unitedutilities.com/resources/files/1342_Map of UU Recreation Sites.pdf (accessed 11 September 2008).

Vigier, François. *Change and Apathy: Liverpool and Manchester during the Industrial Revolution.* Cambridge, MA: MIT Press, 1970.

Wainwright, Alfred. *A Pictorial Guide to the Lakeland Fells.* Bk. 3, *The Central Fells.* 1958. Reprint, London: Michael Joseph, 1992.

[Walker, Adam]. *A Tour from London to the Lakes: Containing Natural, Oeconomical, and Literary Observations, Made in the Summer of 1791. By a Gentleman.* London: John Abraham, 1792.

Walker, David, and Kerry Walker. *Wordsworth and Coleridge: Tour of the Lake District, 1799.* Blackpool: David Walker, 1997.

Ward, James Clifton. *The Geology of the Northern Part of the English Lake District.* London, HMSO, 1876.

Water for Manchester from Thirlmere. The Manchester and Thirlmere Scheme: An Appeal to the Public on the Facts of the Case. Windermere: J. Garnett, 1877.

Watson, John. *The English Lake District Fisheries.* London: Lawrence and Bullen, 1899.

Welch, Thomas V. *How Niagara Was Made Free: The Passage of the Niagara Reservations Act in 1885.* Buffalo: Niagara Frontier Historical Society, 1902.

West, Thomas. *A Guide to the Lakes in Cumberland, Westmorland and Lancashire.* 1784. Reprint, Oxford: Woodstock Books, 1989.

"Westmorland." *Quarterly Review* 122 (1867): 347–81.

Whellan, William. *The History and Topography of the Counties of Cumberland and Westmoreland.* Pontefract: W. Whellan, 1860.

Williams, W. H. *The Commons, Open Spaces, and Footpaths Preservation Society, 1865–1965: A Short History of the Society and Its Work.* London: Commons, Open Spaces, and Footpaths Preservation Society, 1965.

Wilson, Alan. "Technology and Municipal Decision-Making: Sanitary Systems in Manchester 1868–1910." PhD diss., Manchester University, 1990.

Wilson, Albert. *The Flora of Westmorland.* Arbroath: T. Buncle, 1938.

Wilson, J. *Thirlmere to Manchester: A Compendious History of the Promotion, Progress, and Construction of the Great Thirlmere Waterworks Scheme.* Ambleside: George Middleton, 1894.

Winchester, Angus J. L. *The Harvest of the Hills: Rural Life in Northern England and the Scottish Borders, 1400–1700.* Edinburgh: Edinburgh University Press, 2000.

Winter, James. *Secure from Rash Assault: Sustaining the Victorian Environment.* Berkeley: University of California Press, 1999.

Wordsworth, William. *A Guide through the District of the Lakes in the North of England.* 1835. Reprint, edited by Ernest de Selincourt. Oxford: Oxford University Press, 1906.

———. *The Illustrated Wordsworth's Guide to the Lakes.* Edited by Peter Bicknell. New York: Congdon and Weed, 1984.

Yalden, Derek. *The History of British Mammals.* London: T. and A. D. Poyser, 1999.

Contemporary Magazines and Newspapers

(full citations are in the footnotes)
Bolton Evening News
Builder
City Jackdaw
Cumberland and Westmorland Herald
Daily Dispatch
Daily News
Daily Telegraph
Evening Chronicle
Evening Standard
Field
Fishing Gazette
Glasgow Herald
Health Journal
Herald
Housekeeper's Magazine, and Family Economist
Illustrated London News
Lake District Herald
Lancashire Daily Post
Liverpool Daily Post
London Gazette
Manchester City News
Manchester Courant
Manchester Courier
Manchester Examiner and Times
Manchester Faces and Places
Manchester Guardian
Manchester Weekly Times
Medical Press and Circular
North Lonsdale Magazine and Furness Miscellany
Northern Rambler
Pall Mall Gazette
Pall Mall Magazine
Penrith Observer
Punch
Saturday Review
La Science Illustrée
Spectator
Sphere
Standard
Times
West Cumberland Times
Westmorland Gazette
Yorkshire Post and Leeds Intelligencer

Archival Resources

Armitt Library, Ambleside, Cumbria
Ephemera Collection
Lake District Collection
British Library, London
George Richard Jesse, "Public Rights of Way" (scrapbook)
Cumbria Record Office, Carlisle
Bibliotheca Jacksoniana
Cumberland County Council, Minutes of Council and Committees
Cumbria Record Office, Kendal
Greater Manchester County Record Office, Manchester
M.C.W.W. Annual Statements (1869 to 1889–90)
Manchester Central Library, Manchester
MIT (Massachusetts Institute of Technology) Archives and Special Collections, Cambridge, MA
John Ripley Freeman Papers

Illustration Credits

Images are listed by page number.

Introduction

viii Some cities and reservoirs. Map by Drew Laughland, GIS Services, MIT Libraries.

Chapter 1

10 Thirlmere, labeled "Wiburn Water." John Ogilby, *The Roads from Kendal* . . . (London: Abel Swall and Robert Morden, 1698).

14 "Thirlmere and Wytheburn." Print by J. B. Pyne/T. Pickering (London: Day and Son, 1859).

15 "Ullswater, Grasmere & District." *Ward and Lock's (late Shaw's) Pictorial and Historical Guide to the English Lakes* (London: Ward, Lock, 1890), following p. 172.

16 Rock of Names. Reproduced by permission of Dove Cottage, the Wordsworth Trust.

17 Using Thirlmere Water. Print by G. Pickering/W. LePetit (London: Fisher, 1833).

Chapter 2

Chapter 3

Chapter 4

Chapter 5

Index

Page numbers in italics refer to illustrations.